MARINE MICROBIAL ECOLOGY

By

Dr. D.R. Khanna

Reader in Zoology
Gurukul Kangri University
Haridwar
(Uttaranchal)
(India)

DISCOVERY PUBLISHING HOUSE PVT. LTD.
NEW DELHI-110 002

Published by:
Namit Wasan

DISCOVERY PUBLISHING HOUSE PVT. LTD.
4383/4B, Ansari Road, Darya Ganj
New Delhi-110 002 (India)
Phone : +91-11-23279245; 23253475; 43596065
E-mail : discoverybooksindia@gmail.com
discoverypublishinghouse@gmail.com
namitwasan9@gmail.com
web : www.discoverypublishinggroup.com

First Edition: **2010**

Reprinted: **2024**

ISBN: 978-81-8356-575-2

Marine Microbial Ecology

Printed at:
Infinity Imaging Systems
Delhi

Preface

Modern marine microbial ecology can be considered to have started in the 1970s, when it was shown that most respiration in the oceans was in the bacterial size fractions and that bacteria ere very abundant. Before that time, microbes were not even considered in the ecology of the oceans. Now-a-days, however, marine microorganisms are nown to be responsible for half of the total primary production on the planet, and the 1030 icrobial cells present in the oceans account for more than 95% of the total respiration. The hange of perspective, therefore, has been spectacular.

The application of genomic approaches to marine microbial ecology in the past few years has caused a kind of Copernican revolution. Thanks to such techniques, novel functions have been discovered, a large diversity of microorganisms has been uncovered, and the meaning of concepts such as pecies, genome, and niche has been challenged This article reviews the different genomics-based approaches relevant for marine microbial ecology and uses one example from each to demonstrate their tremendous possibilities.

The Microbial Habitat Group investigates the structure, dynamics and evolution of microbial habitats. This includes research on the physical and chemical characteristics of microbial habitats, such as transport processes between marine sediments and the water column, and energy fluxes across oxic/anoxic hemoclines, between light and dark environments and from solid to solute and gaseous phases. Together with the observation of physico-chemical characteristics we analyze their ecological consequences, and develop models for these processes. Our research is accompanied by advances in marine technologies to resolve the spatial and temporal dynamics of microbial habitats in situ as well as in experimental set-ups.

Among the microbial habitats currently investigated are detritus particles, coral exudates, sandy sediments, intertidal flats, shallow- and deep-water hydrocarbon seeps, mud volcanoes, gas hydrates. Our research is tightly linked to the characterization of the microbial populations dominating individual habitats (by the department of Molecular Ecology and Microbiology), as well as to the analysis of their function and contribution to carbon flow (by the Biogeochemistry and Microsensor group).

The microbial habitat describes as the physical location and type of environment in which a population of microorganisms lives. Hence, this research group studies the physical, chemical, geological hydrological and biological characteristics of distinct microbial habitats. The goal of our research is to understand structure and change of microbial ecosystems, the formation of niches for microbial populations and to investigate environmental dynamics and their consequences on the occurrence, biodiversity and distribution of microbial populations. The uniting topics to all researchers in the group is to obtain: (1) "true" quantitative insight to ecosystem structure, dynamics and biogeochemical fluxes based on *in-situ* measurements; and (2) insight into the related variations in microbial biodiversity on relevant spatial and temporal scales.

– Author

Contents

1

Introduction

Genomics has brought about a revolution in all fields of biology. Before the development of microbial ecology in the 1970s, microbes were not even considered in marine ecological studies. Today we know that half of the total primary production of the planet must be credited to micro-organisms. This and other discoveries have changed dramatically the perspective and the focus of marine microbial ecology.

The application of genomics-based approaches has provided new challenges and has allowed the discovery of novel functions, an appreciation of the great diversity of micro-organisms, and the introduction of controversial ideas regarding the concepts of species, genome, and niche. Nevertheless, thorough knowledge of the traditional disciplines of biology is necessary to explore the possibilities arising from these new insights. This work reviews the different genomic techniques that can be applied to marine microbial ecology, including both sequencing of the complete genomes of micro-organisms and metagenomics,which, in turn, can be complemented with the study of mRNAs (transcriptomics)and proteins (proteomics).

The example of proteorhodopsin illustrates the type of information that can be gained from these approaches. A genomics perspective constitutes a map that will allow microbiologists to focus their research on potentially more productive aspects.

Neither genomics nor marine microbial ecology existed at the time of publication of *The Microbe's Contribution to Biology*. However, there is a parallel between the book's content and the current role of marine genomics: in both cases, the deeper knowledge of microbes served as an eye opener that revolutionized our understanding of life's possibilities. In both cases, there was an appreciation that all the disciplines of biology (biochemistry, genetics, physiology, taxonomy, etc.) would be simultaneously needed in order to exploit the knowledge that could be obtained by these new approaches.

The biological sciences are experiencing a revolution, both technical and intellectual, in good part due to the legacy left by the human genome sequencing project, for example, the fact that biology is being transformed from a datapoor into a data-rich science. The ever-growing number of sequences allows biologists to ask new questions and to approach the old ones from a more-informed perspective. The questions that can be formulated now go beyond pure genetics (in the sense of genome organisation or diversity and the phylogeny of life) and include both the functioning of living beings as a whole and their adaptation to the environment. In fact, getting sequences is no longer a problem.

Modern marine microbial ecology can be considered to have started in the 1970s, when it was shown that most respiration in the oceans was in the bacterial size fractions and that bacteria ere very abundant. Before that time, microbes were not even considered in the ecology of the oceans. Now-a-days, however, marine microorganisms are nown to be responsible for half of the total primary production on the planet , and the 1030 icrobial cells present in the oceans account for more than 95% of the total respiration The hange of perspective, therefore, has been spectacular.

The application of genomic approaches to marine microbial ecology in the past few years has caused a kind of Copernican revolution. Thanks to such techniques, novel functions have been discovered, a large diversity of microorganisms has been uncovered, and the meaning of concepts such as pecies, genome, and niche has been challenged This article reviews the different genomics-based approaches relevant for marine microbial ecology and uses one example from each to demonstrate their tremendous possibilities.

From Genes to Genomes: A Change of Scale

Certain messenger RNAs can be analyzed by reverse transcription to DNA and subsequent cloning. In this case, the technique will confirm not only that the corresponding gene is there, but also that the gene is being expressed actively.

Finally, particular proteins can be searched for. This is more difficult to do in natural samples, because proteins cannot be amplified by PCR as nucleic acids can. However, if a protein is sufficiently abundant, there is no reason why it cannot be purified from the environment. This would idence that the gene is present, that it is being expressed, and that the corresponding protein is being synthesized at the particular place and time where the sample was originally taken.

The equivalent approaches when a micro-organism has been previously isolated in axenic culture. In this case, all the techniques are easier to carry out, because the organism can be grown to high concentrations, and because most genes will be represented by a single copy (since there will be a single species). The problem is that isolation in axenic culture is a selective process that only retrieves some of the microbes that are important in nature.

The study of genomes (genomics in the case of single microbial species or metagenomics in the case of whole communities) can be complemented with the study of the mRNAs (transcriptome, ranscriptomics) or proteins (proteome, proteomics). Obviously, this change of scale requires two hings: a large sequencing capacity and very powerful bioinformatics tools to organize and study the massive amount of information. Both these things received a decisive impulse with the human genome sequencing project, and the sequencing capabilities and bioinformatics resources that resulted from it are now widely available.

GENOMIC APPROACHES TO MARINE MICROBIAL ECOLOGY

This section will review the different approaches available to study DNA (genomics and metagenomics). Consideration of either ranscriptomics or proteomics exceeds the scope of the present paper. Some approaches to study DNA are summarized.

Genomics of Marine Microorganisms Isolated in Axenic Culture

Until one year ago, most sequenced genomes belonged to bacteria of medical interest. However, thanks to the initiative of the Gordon and Betty Moore Foundation, over 150 genomes of marine bacteria have been sequenced in the last two years. Of course, the main caveat is that many of the bacteria isolated in axenic culture are not the most important in nature.

However, these genomes offer a wealth of information that can be used to interpret the results from transcriptomics and proteomics. For example, the complete genomes of four cyanobacteria have ffered novel insights into the meaning of concepts such as species and ecological niche.

Marine cyanobacteria are grouped into two closely related genera according to their 16S rRNA: *rochlorococcus* and *Synechococcus*. It is estimated that these cyanobacteria are responsible for two-thirds of the photosynthesis in the oceans, and they are widely distributed in the equatorial and temperate latitudes of all oceans. In fact, *rochlorococcus* may be the most abundant photosynthetic organism on Earth. Recently, the genomes of four different strains were sequenced: One from *Synechococcus* and three from *Prochlorococcus*. According to conventional 16S rRNA criteria, the three *Prochlorococcus* strains belonged to the same species.

However, the genome of strain MIT9313 turned out to have 2.4 megabases (Mb), while those of the other two strains (MED4 and SS120) had only 1.7 Mb. It was very shocking to realize that organisms belonging to the same "species" may ve genomes with such different sizes. Looking at the natural distribution of each strain, it could be appreciated that the size of the genome seemed to be related to the stability of the particular train's environment.

Thus, strain MED4 is adapted to live at the surface of the ocean, with relatively constant conditions of high light intensity and low nutrient availability, while strain SS120 is adapted to lower depths, where low light intensities and higher nutrient conditions are relatively constant. Strain MIT9313 inhabits intermediate depths, where conditions are likely to be more variable. The case of the *Synechococcus* strain is similar. It also has a 2.4-Mb genome and lives in upwelling and vertical mixing situations, where conditions are also more variable. Apparently, the two latter strains require n additional 0.7 Mb in their genomes to have enough versatility to survive in their changing environments. In effect, when genes conferring the ability to use different nitrogen sources were searched for in the four genomes, all were found to contain genes involved in the metabolism of ammonia. However, while the two strains with small genomes could only use one or two additional compounds, the two with a large genome could use numerous sources of nitrogen, such as nitrate, nitrite, urea, amino acids, peptides, and even cyanate, a substance that had not been known to act as a nitrogen source for any microorganism.

It therefore seems that the two strains adapted to constant environments discarded all the genes that were not strictly necessary for their growth, a phenomenon described as "genome streamlining". It is intriguing that evolution allows such major changes that affect full blocks of genes but which retain very similar 16S rRNA. In summary, the example of marine cyanobacteria demonstrates how genomic studies may dramatically alter our understanding of the taxonomy, evolution, physiology, and ecology of microorganisms.

Metagenomics

In one metagenomics approach, large fragments of environmental DNA can be carefully extracted and cloned in appropriate vectors, such as fosmids or BAC libraries. These large fragments of DNA (up to 100 kb) contain several genes arranged in the precise order in which they were found in the genome they came from. If a gene for 16S rRNA is found in one clone, the bacterium it came from can be identified and the neighbouring genes can be sequenced. In this way, the existence of a novel function was discovered in an uncultivated bacte ium. A clone from one such library had the 16S rRNA gene of SAR86 (a cluster of sequences retrieved from many oceans but with no representative in axenic culture). A gene coding for a protein imilar to halorhodopsin was found in the same clone. Upon further study, this protein (named proteorhodopsin) was shown to use light to generate a proton gradient across the cell membrane; thus, an unknown group of bacteria, SAR86, was shown to have a novel function (phototrophy) An alternative metagenomics approach involves fragme ting the environmental DNA into small fragments around 3 kb) and cloning those fragments in conventional vectors in a procedure called "shotgun cloning."

Nonetheless, by shotgun cloning a sample from the Sargasso Sea, Venter et al. revealed over 00,000 genes, many of them with unknown functions. In one of the most typical examples, this strategy has increased by at least one order of magnitude the number of roteorhodopsin genes known from previous approaches.

Accordingly, DNA fragments can be sequenced without previous screening. This allows for the discovery of novel genes regardless of their origin. The drawbacks are that this method requires massive sequencing to analyze the thousands of clones generated from a single sample, and that it is difficult to reassemble genomes from many small fragments.

New sequencing techniques are making metagenomic approaches cheaper and, therefore, accessible to most laboratories. The sequencing technology offered by the 454 Life Sciences company, among others, is being used with increasing frequency. Although the sequenced fragments are, for the time being, very short (around 100 bases long) and this entails difficulties in the assembly of the fragments, the technique has been successfully used to sequence an entire bacterial genome in one week A third option is to use PCR with primers specific for a given group of bacteria or for a given gene, for example 16S rRNA. Then, the collection of sequences amplified can be subject to the same shotgun cloning and massive sequencing described above. Thus, Sogin et al. have used this strategy, in combination with 454 sequencing, to screen deepocean samples for 16S rRNA genes. The addition of an amplification step and the enormous number of sequences that were subsequently generated have incr ased the known diversity of marine samples by one or two orders of magnitude over what was previously recognized from conventional cloning and sequencing. The large number of different sequences in one sample is in agreement with the idea that bacterial communities are formed by a few dozen abundant taxa and a very large collection of rare taxa, the former making up the diversity of that ecosystem and the latter contributing to its complete biodiversity Finally, if metagenomics libraries from two or more different communities are available, they can be used to compare the relative abundance of genes with given functions and these can be related back to the particular conditions in each environment.

That genes for chemotaxis were over-represented in the whale fall communities. This suggests that chemotaxis is important for the bacteria to find and move towards the episodic whale fall in the bottom of the ocean. At the surface of the Sargasso Sea, by contrast, resources may be more regularly mixed by turbulence, making hemotaxis a less useful property. In any case, this comparison immediately suggests interesting ecological hypotheses that can be tested in further studies.

The Return to General Microbiology: The Case of Proteorhodopsin

A most welcome consequence of the revolution rought forward by genomics is the need to recover all the traditional disciplines of biology. In order to make sense out of the millions of basepairs of sequence, it is necessary to go back to the biochemistry textbooks to correctly interpret the metabolic pathways that can be reconstructed from genomes. It is also necessary to isolate more organisms in axenic culture and to characterize them with proper and careful tax nomy. And, of course, the tools and concepts of genetics are essential to try to understand how whole genomes function. Some of these aspects can be illustrated with the example of proteorhodopsins.

As mentioned earlier, proteorhodopsin (PR) was discovered through the large-fragment metagenomics approach. Small-fragment metagenomics increased by an order of magnitude the number of genes known to code for PRs found in natural samples. Up to this point, however, all that was available was the DNA sequence of one gene. But was this gene expressed? Did it code for active proteins? Did the protein confer any advantage to the bacteriapossessing it? Some of these questions could be answered by cloning the PR gene in Escherichia coli Luckily, E. coli did express the gene and was able to put the protein to use. Upon illumination, the clones generated a proton gradient and this was only the case if light had the wavelength absorbed by retinal, the pigment in PR. These biochemical and physiological studies would have been impossible if only the sequence were available. It was necessary to have the gene within a living cell, in this case an *E. coli* cell. However, in order to determine whether marine bacteria could use PR for faster or more efficient growth, axenic cultures of bacteria naturally having the gene were needed.

After isolation in axenic culture and sequencing of the whole genome, the Pelagibacter ubique, one of the most abundant bacteria in the surface oceans, had the PR gene. With an axenic culture of a PR-containing bacterium, it became possible to do the biochemical and physiological studies carried out in *E. coli* in a more natural environment. Furthermore, the effects of PR on the growth of the organism could be analyzed. Giovannoni et al. detected the formation of a proton gradient upon illumination with the appropriate wavelengths. But they could not show any difference in the growth of the organisms in the light or in the dark. Thus, the role of PR in natural bacteria remained a mystery.

At the Institut de Ciencies del Mar, CSIC, in Barcelona, a large collection of bacteria had been isolated in axenic culture from the Blanes Bay Microbial Observatory Nine of these bacteria were selected for the genome sequencing initiative of the Gordon and Betty Moore Foundation and several genomes became available at the end of 2005. Upon analysis of their genomes, the presence of carotenoid genes was detected in all three isolates.

This was expected, since the isolates were either yellow or orange. MED217 and MED134 had the genes crt(EBIY), necessary for the synthesis of *f*β-carotene, and crtZ, necessary to convert *f*β-carotene into zeaxanthin. In addition, strain MED152 had the genes crtD and crtA as well as an additional copy of crtZ. Since the axenic cultures were available, we could examine whether the two former strains were yellow and, using HPLC, whether the only carotenoids ere *f*β-carotene and zeaxanthin. Strain MED152 was orange and had an unidentified carotenoid in addition to *f*β-carotene and zeaxanthin.

This was obviously synthesized by the products of genes crtD and crtA, but the identity of this compound is still unknown.

The analysis of the three genomes also showed that strains MED134 and MED152 had two additional genes related to light: blh, a gene necessary to synthesize retinal from *f*β-carotene, and the opsin gene. Thus, both the protein and the pigment that make up PR could be synthesized by these two strains. Thanks to the fact that axenic cultures are available, we can now check whether these genes are expressed in vivo. Preliminary experiments show promise to clarify the role of PR in natural marine bacterioplankton. The finding of PR in marine bacteria indicates that a reinterpretation of bacterial heterotrophic activity in the sea may be in order. Measurements of this parameter have been traditionally carried out in the dark. If a substantial fraction of the community carries PR genes, however, measurements of carbon flow through bacteria may have been significantly underestimated. Out of the same amount of available dissolved organic matter, a microbial community rich in PR will produce more particulate organic matter than a community poor in PR, thus increasing the efficiency of carbon transfer in the microbial food web. I think this example nicely illustrates the rich interactions between genomics on the one hand and the traditional disciplines of biology on the other.

The Microbial Habitat Group investigates the structure, dynamics and evolution of microbial habitats. This includes research on the physical and chemical characteristics of microbial habitats, such as transport processes between marine sediments and the water column, and energy fluxes across oxic/anoxic hemoclines, between light and dark environments and from solid to solute and gaseous phases. Together with the observation of physico-chemical characteristics we analyze their ecological consequences, and develop models for these processes. Our research is accompanied by advances in marine technologies to resolve the spatial and temporal dynamics of microbial habitats in situ as well as in experimental set-ups.

Among the microbial habitats currently investigated are detritus particles, coral exudates, sandy sediments, intertidal flats, shallow- and deep-water hydrocarbon seeps, mud volcanoes, gas hydrates. Our research is tightly linked to the characterization of the microbial populations dominating individual habitats (by the department of Molecular Ecology and Microbiology), as well as to the analysis of their function and contribution to carbon flow (by the Biogeochemistry and Microsensor group).

Main Themes of Scientific Work

The microbial habitat describes the physical location and type of environment in which a population of microorganisms lives. Hence, this research group studies the physical, chemical, geological hydrological and biological characteristics of distinct microbial habitats. The goal of our research is to understand structure and change of microbial ecosystems, the formation of niches for microbial populations and to investigate environmental dynamics and their consequences on the occurrence, biodiversity and distribution of microbial populations. The uniting topics to all researchers in the group is to obtain:

1. "true" quantitative insight to ecosystem structure, dynamics and biogeochemical fluxes based on *in-situ* measurements; and
2. insight into the related variations in microbial biodiversity on relevant spatial and temporal scales.

The development of novel instrumentation for *in-situ* studies of submarine ecosystems, ranging from coastal sands to reefs, continental margins, polar waters and hydrothermal vents enables the Habitat group in collaboration with the Microsensor group to improve the quantification of transport and reaction in these habitats, which are dominant factors structuring microbial habitats. Furthermore, we link our in situ biogeochemistry and biodiversity studies closely to the investigation of microbial function in the respective habitats, in collaboration with the department of Microbiology and Molecular Ecology.

Coastal Sands as Microbial Habitats

The coastal zones of the ocean are highly productive ecosystems, and play an important role in the global cycles of carbon and nitrogen. It has been estimated that about 30% of the continental shelf sediments in up to 40 m water depth receive enough light for benthic photosynthesis. Bacterial activity and biodiversity in sandy sediments of our main study site at the island of Sylt have been shown to be extremely high and closely correlated with the distribution of microalgae within the sediment. A major part of the research on coastal biogeochemical processes was performed within the COSA (EU; ended in 2005) and WATT (DFG) projects (currently in its second phase). As a follow up to COSA, current projects on coastal habitats include: (1) the investigation of spatial and temporal changes in abiotic and biotic environmental parameters and their correlation with dynamics in microbial populations; (2) the effects of advection on benthic photosynthesis; (3) mechanistic studies of advection at the scale of individual sand grains. The potential contribution of deeper sub-tidal benthic photosynthesis to the coastal carbon cycle is investigated with a range of in situ instruments including eddy correlation and benthic crawlers equipped with microelectrodes, planar optodes, benthic chambers, and surface scanners. Progress has been made in the past 2 years with: quantifying deep porewater circulation and nutrient efflux from tidal flats finding very high benthic primary production in advective sublittoral sands compared to muds revealing ecological relationships between microbial biodiversity in sandy sediments and benthic productivity testing the effect of dwelling macrofauna on bacterial distribution and diversity.

Focused flow of hydrocarbon gases, fluids and muds rising from great depth to surface sediments, hydrothermal discharge as well as episodic depositions of organic carbon from dead marine mammals, woods and kelps are a spatially limited source of energy to the food-impoverished deep sea benthos. We investigate how specialized communities of microorganisms utilize these energy sources, and provide the basis for highly productive benthic ecosystems. A main focus is the structure and functioning of gas escape pathways in the ocean and the special ecosystems connected to them. Our research on microbial methane turnover and cold seep ecosystems is supported by external funding from the European Union "HERMES", and the BMBF/DFG programme Geotechnologien "Methane in the Geo-Biosphere". Furthermore we are actively involved in the project area E on fluid and gas seepage of the DFG research center on ocean margins (RCOM/MARUM), recently awarded an excellence cluster funding. In collaboration with the Microbial Symbiosis group, transport processes and energy fluxes at different hydrothermal vents are investigated within the frame of the DFG SPP1144 "From mantel to ocean: energy, geochemistry and life cycles at spreading centers", and in the future also within the excellence cluster MARUM. Within the MPG-CNRS GDRE DIWOOD and the ESF EUROCORES EuroDeep project CHEMECO we have started research on the microbial ecology of large food falls like woods and whale carcasses, to investigate microbial degradation processes forming anoxic habitats in the deep sea.

A first global data set on *in-situ* respiration rates of oxygen and sulfate at hydrocarbon seeps flow through incubations to establish methane oxidation kinetics and methane assimilation rates comparative analysis of methane emission and consumption at different European mud volcano systems and the relationship to microbial community structure investigation of the effect of natural CO_2 leakage on the distribution of bacteria and archaea in seep sediments investigation of sulfur fluxes in diffuse flow dominated microbial habitats at vents understanding biodiversity patterns of bacteria in relation to methane flux and sediment bioturbation at cold seeps.

Microbes occupy distinct niches on animal surfaces, exudates and tissues forming symbiotic, commensalistic or pathogenic relationships with their hosts. An interesting question remains if deep-water animal biodiversity hotspots on continental margins like cold-water coral reefs and associated sponge accumulations also provide distinct microbial habitats. The microbial community structure, diversity and dynamics of the different reef-associated microbial habitats such as coral mucus, coral skeleton surface and coral tissue, as well as proximal sediment and ambient seawater are being investigated using high-resolution molecular techniques. Special focus will be given to the question in how far cold-water corals act as "ecosystem engineers" for microbial communities by shaping their diversity via habitat formation or release of organic matter.

Sponges are evolutionary ancient marine organisms, which may host high amounts of associated microbes ("bacterio-sponge"). The aim of our studies on sponges as microbial habitats is to understand the impact of microbial processes on the metabolic capacities of the sponge animal as a whole, with a special focus on anaerobic processes of the nitrogen and sulfur cycle. Main achievements during the last two years include:

- Distinct microbial communities could be found in the different coral reef habitats (coral mucus, coral surface, seawater and sediment) both in situ and ex situ (Master thesis Sandra Schöttner).
- Species-specific microbial communities could be detected in different sponge species from cold-water coral reefs (preliminary results).
- The detailed investigation of anoxic tissue zones in many sponge species both *in-situ* and *ex-situ*.
- First quantification of anaerobic microbial processes in sponges including sulphate reduction, denitrification and anammox.

Methods Developments: *In-situ* Technologies for Microbial Habitat Studies

Microbial habitats characterized by intermediate to high fluid advection, like cold seeps, hydrothermal vents, deep-water coral reefs, and coastal sands, are shaped by a complex interplay of physical, biological, geochemical, and geological processes. Often biogeochemical and physicochemical gradients are extremely steep and spatially and temporally variable in these ecosystems. Hence, to quantify chemical gradients, microbial processes and transport rates we develop and operate *in-situ* instruments in collaboration with the microsensor group and the electronical and mechanical workshops. For high-resolution chemical gradients we use the microprofiler and/or planar optode module, to measure oxygen exchange and oxidation rates we operate EDDY, benthic chambers, and INSINC, and for hydrographical and physical parameters we utilize Lance-A-Lot and DeepFlow. These tools are deployed from ships, by remotely operating underwater vehicles (ROV and crawlers) and submersibles, which allow targeted sampling and a precise positioning of instruments at selected spots under visual control. Adaptations include miniaturizing of the measuring devices as well as improvement of the operability such as multiple uses during one dive and long-term measurements, and are carried out in collaboration with leading international research institutions operating ROV's, submersibles and benthic crawlers.

These *in-situ* technology achievements now lead into the development of deep-sea observatories for long-term ecosystem studies (EU-NoE ESONET). Main achievements in the past two years include:

- routine utilization of the in situ incubator INSINC for sulfate reduction rates, and of EDDY for oxygen fluxes in coastal sediments.
- repetitive microsensor profiling (up to 80 measurements during one dive).
- horizontal profiling of microbial reefs in the Black Sea.
- Integration of *in-situ* modules for measurement of coastal productivity and carbon flux to the benthic crawler MOVE (Marum, Bremen).
- High resolution oxygen concentration measurements by optical fiber in oligotrophic deep subsurface cores design of a bottom landing elevator for flexible combinations of ROV payloads.

Method Developments: Biodiversity Patterns in Dynamic Microbial Habitats

One key ecological parameter for the comparison of habitats is their biodiversity, i.e. the species richness and the community structure. In microbial ecology, simple questions e.g. as to the relative diversity and abundance of Bacteria vs Archaea in different marine habitats, or the existence of biogeographical patterns in communities or specific taxonomic groups have not been answered yet. Reproducible high-throughput, high-resolution molecular methods for the assessment of biodiversity are now available, but often knowledge is missing about environmental dynamics on relevant spatial and temporal scales. We have adapted methods for the comparative assessment of biodiversity in different biogeochemically well investigated microbial habitats, including coastal sandy sediments, coral-associated surfaces, microbial mats, or deep-sea sediments impacted by high hydrocarbon fluxes. Changes in microbial community structure are examined by using a combination of cutting-edge, highly reproducible molecular ecology techniques such as Terminal Restriction Fragment Length Polymorphism (T-RFLP), Automated Ribosomal Intergenic Spacer Analysis (ARISA) and 454 massive tag sequencing. Those techniques provide a thorough analysis of the source of community variation among numerous samples. By combined analysis of the assessed biodiversity and biogeochemical parameters, the effects of the main factors affecting microbial community structure can be quantified, as well as the amount of covariation and interactions among the factors. Main achievements in the last two years are:

Development of a toolbox for high throughput community structure analysis (e.g. T-RFLP, ARISA, 454 tag sequencing) that enables exploratory analyses of a huge number of samples. This is a key step to identify samples whose community structures are particularly interesting to be further investigated for phylogeny and function, which is not possible by traditional clone library approaches.

Comparison of traditional community fingerprinting techniques (e.g. T-RFLP, ARISA) with more advanced technique (e.g. 454 tag sequencing) to describe community structure and validation of the tools to describe microbial ecology in coastal ecosystems.

Integration of biological, environmental, spatial and temporal data into coherent statistical and ecological frameworks. This has helped identify the main factors structuring biodiversity in various marine ecosystems.

2

Seawater

INTRODUCTION

Seawater is water from a sea or ocean. On average, seawater in the world's oceans has a salinity of about 3.5%, or 35 parts per thousand (also expressed 35% or 35 ppt). This means that every 1 kg of seawater has approximately 35 grams of dissolved salts (mostly, but not entirely, the ions of sodium chloride: Na^+, Cl^-). The average density of seawater at the surface of the ocean is 1.025 g/ml; seawater is denser than fresh water (which reaches a maximum density of 1.000 g/ml at a temperature of 4°C) because of the added weight of the salts and electrostriction. The freezing point of sea water decreases with increasing salinity and is about -2°C (28.4°F) at 35 parts per thousand.

Although the vast majority of seawater has a salinity of between 3.1% and 3.8%, seawater is not uniformly saline throughout the world. Where mixing occurs with fresh water runoff from river mouths or near melting glaciers, seawater can be substantially less saline. The most saline open sea is the Red Sea, where high rates of evaporation, low precipitation and river inflow, and confined circulation result in the formation of unusually salty seawater. The salinity in isolated bodies of water (for example, the Dead Sea) can be considerably greater.

The density of surface seawater ranges from about 1020 to 1029 $kg{\cdot}m^{-3}$, depending on the temperature and salinity. Deep in the ocean, under high pressure, seawater can reach a density of 1050 kg·m^{-3} or higher. Seawater pH is limited to the range 7.5 to 8.4. The speed of sound in seawater is about 1500 m·s^{-1}, and varies with water temperature and pressure.

COMPOSITIONAL DIFFERENCES FROM FRESH WATER

Seawater is more enriched in dissolved ions of all types than fresh water. However, the ratios of various solutes differ dramatically. For instance, although seawater is about 2.8 times more enriched with bicarbonate than river water based on molarity, the percentage of bicarbonate in seawater as a ratio of all dissolved ions is far lower than in river water; bicarbonate ions constitute 48% of river water solutes, but only 0.41% of all seawater ions. Differences like these are due to the varying residence times of seawater solutes; sodium and chlorine have very long residence times, while calcium (vital for carbonate formation) tends to precipitate out much more quickly.

GEOCHEMICAL EXPLANATIONS

Table 2.1: Total Molar Composition of Seawater (Salinity = 35)

Component	Concentration (mol/kg)
H_2O	53.6
Cl^-	0.546
Na^+	0.469
Mg^{2+}	0.0528
SO_4^{2-}	0.0282
Ca^{2+}	0.0103
K^+	0.0102
C_T	0.00206
Br^-	0.000844
B_T	0.000416
Sr^{2+}	0.000091
F^-	0.000068

Scientific theories behind the origins of sea salt started with Sir Edmond Halley in 1715, who proposed that salt and other minerals were carried into the sea by rivers, having been leached out of the ground by rainfall runoff. Upon reaching the ocean, these salts would be retained and concentrated as the process of evaporation removed the water. Halley noted that of the small number of lakes in the world without ocean outlets (such as the Dead Sea and the Caspian Sea), most have high salt content. Halley termed this process "continental weathering".

Halley's theory is partly correct. In addition, sodium was leached out of the ocean floor when the oceans first formed. The presence of the other dominant ion of salt, chloride, results from "outgassing" of chloride (as hydrochloric acid) with other gases from Earth's interior via volcanos and hydrothermal vents. The sodium and chloride ions subsequently became the most abundant constituents of sea salt.

Ocean salinity has been stable for billions of years, most likely as a consequence of a chemical/tectonic system which removes as much salt as is deposited; for instance, sodium and chloride sinks include evaporite deposits, pore water burial, and reactions with seafloor basalts Since the ocean's formation, sodium is no longer leached out of the ocean floor, but instead is captured in sedimentary layers covering the bed of the ocean. One theory is that plate tectonics result in salt being forced under the continental land masses, where it is again slowly leached to the surface.

HUMAN CONSUMPTION OF SEAWATER

Accidentally consuming small quantities of clean seawater is not harmful, especially if the seawater is consumed along with a larger quantity of fresh water. However, consuming seawater to maintain hydration is counterproductive; in the long run, more water must be expended to eliminate the seawater's salt (through excretion in urine) than the amount of water that is gained from drinking the seawater itself.

This occurs because the amount of sodium chloride in human blood is actively regulated within a very narrow range of 9 g/L (0.9% by weight) by the kidney. Drinking seawater (which contains about 3.5% ions of dissolved sodium chloride) temporarily increases the concentration of sodium chloride in the blood. This in turn promotes sodium excretion by the kidney, but the sodium concentration of seawater is above the maximum concentrating ability of the human kidney. Eventually with further seawater intake the blood concentration of sodium will rise to toxic levels, removing water from all cells and interfering with nerve conduction ultimately giving seizures and heart arrhythmias which become fatal. Of note, various animals adapt to harsh living conditions. For example, the desert rat is able to concentrate sodium far more efficiently than the human kidney, and therefore would be able to survive by drinking seawater.

Survival manuals consistently advise against drinking seawater. For example, the book "Medical Aspects of Harsh Environments" presents a summary of 163 life raft voyages. The risk of death was 39% for those who drank seawater, compared to only 3% for those who did not drink seawater. The effect of seawater intake has also been studied in laboratory setting in rats. (Etzion and Yagil; *Metabolic effects in rats drinking increasing concentrations of seawater*. Comp Biochem Physiol A. 1987;86(1):49-55.) This study confirmed the negative effects of drinking seawater when dehydrated.

The temptation to drink seawater has always been greatest for sailors who have expended their supply of fresh water, and are unable to capture enough rainwater for drinking. This frustration is described famously by a line from Samuel Taylor Coleridge's *The Rime of the Ancient Mariner*:

"Water, water, every where,

And all the boards did shrink water, water every where,

Nor any drop to drink."

Although it is clear that a human cannot survive on seawater alone, some people claim that one can drink up to two cups a day, mixed with fresh water in a 2:3 ratio, without ill-effect. The French physician Alain Bombard claimed to have survived an ocean crossing in a small raft using only seawater and other provisions harvested from the ocean, but the veracity of his findings was challenged. In Kon-Tiki, Thor Heyerdahl reported drinking seawater mixed with fresh in a 40 : 60% ratio. A few years later another adventurer named William Willis claimed to have drunk two cups of seawater and one cup of fresh per day for 70 days without ill-effect when he lost his water supply.

Most modern ocean-going vessels create drinkable (potable) water from seawater using desalination processes such as vacuum distillation, multi-stage flash distillation, or by the use of reverse osmosis. However these processes are energy intensive, and most were not available or practical during the age of sail.

SALINITY

Salinity is the saltiness or dissolved salt content of a body of water. Salinity in Australian English and North American English may also refer to the salt in soil.

The technical term for saltiness in the ocean is halinity, from the fact that halides - chloride specifically - are the most abundant anions in the mix of dissolved elements. In oceanography, it has been traditional to express salinity not as percent, but as parts per thousand (ppt or ‰), which is approximately grams of salt per litre of solution. Other disciplines use chemical analyses of solutions, and thus salinity is frequently reported in mg/L or ppm (parts per million). Prior to 1978, salinity or halinity was expressed as ‰ usually based on the electrical conductivity ratio of the sample to "Copenhagen water", an artificial sea water manufactured to serve as a world "standard". In 1978, oceanographers redefined salinity in the Practical Salinity Scale (PSS) as the conductivity ratio of a sea water sample to a standard KCl solution. Ratios have no units, so it is not the case that a salinity of 35 exactly equals 35 grams of salt per litre of solution.

These seemingly esoteric approaches to measuring and reporting salt concentrations may appear to obscure their practical use; but it must be remembered that salinity is the sum weight of many different elements within a given volume of water. It has always been the case that to get a precise salinity as a concentration and convert this to an amount of substance (sodium chloride, for instance) required knowing much more about the sample and the measurement than just the weight of the solids upon evaporation (one method of determining "salinity"). For example, volume is influenced by water temperature; and the composition of the salts is not a constant (although generally very much the same throughout the world ocean). Saline waters from inland seas can have a composition that differs from that of the ocean. For the latter reason, these waters are termed saline as differentiated from ocean waters, where the term haline applies (although is not universally used).

Marine waters are those of the ocean, another term for which is euhaline seas. The salinity of euhaline seas is 30 to 35. Brackish seas or waters have salinity in the range of 0.5 to 29 and metahaline seas from 36 to 40. These waters are all regarded as thalassic because their salinity is derived from the ocean and defined as homoiohaline if salinity does not vary much over time (essentially invariant). The table on the right, modified from Por (1972) follows the "Venice system" (1959).

In contrast to homoiohaline environments are certain poikilohaline environments (which may also be thallassic) in which the salinity variation is biologically significant Poikilohaline water salinities may range anywhere from 0.5 to greater than 300. The important characteristic is that these waters tend to vary in salinity over some biologically meaningful range seasonally or on some other roughly comparable time scale. Put simply, these are bodies of water with quite variable salinity.

Highly saline water, from which salts crystallize (or are about to), is referred to as brine.

Environmental Considerations

Salinity is an ecological factor of considerable importance, influencing the types of organisms that live in a body of water. As well, salinity influences the kinds of plants that will grow either in a water body, or on land fed by a water (or by a groundwater). A plant adapted to saline conditions is called a halophyte. Organisms (mostly bacteria) that can live in very salty conditions are classified as extremophiles, halophiles specifically. An organism that can withstand a wide range of salinities is euryhaline.

Salt is difficult to remove from water, and salt content is an important factor in water use (such as potability).

pH

pH is a measure of the acidity or basicity of a solution. It is defined as the cologarithm of the activity of dissolved hydrogen ions (H^+). Hydrogen ion activity coefficients cannot be measured experimentally, so they are based on theoretical calculations. The pH scale is not an absolute scale; it is relative to a set of standard solutions whose pH is established by international agreement. The concept of pH was first introduced by Danish chemist Søren Peder Lauritz Sørensen at the Carlsberg Laboratory in 1909. Sørensen suggested the notation "PH" for convenience, standing for "power of hydrogen" using the cologarithm of the concentration of hydrogen ions in solution, p[H] Although this definition has been superseded p[H] can be measured if an electrode is calibrated with solution of known hydrogen ion concentration.

Pure water is said to be neutral. The pH for pure water at 25 °C is close to 7.0. Solutions with a pH less than 7 are said to be acidic and solutions with a pH greater than 7 are said to be basic or alkaline. pH measurements are important for medicine, biology, chemistry, food science, environmental science, oceanography and many other applications. pH is defined as minus the decimal logarithm of the hydrogen ion activity in an aqueous solution. By virtue of its logarithmic nature, pH is a dimensionless quantity.

$$\mathrm{pH} = -\log_{10} a_{\mathrm{H}} = \log_{10} \frac{1}{a_{\mathrm{H}}}$$

where aH is the (dimensionless) activity of hydrogen ions. The reason for this definition is that a_H is a property of a single ion which can only be measured experimentally by means of an ion-selective electrode which responds, according to the Nernst equation, to hydogen ion activity. pH is commonly measured by means of a combined glass electrode, which measures the potential difference, or electromotive force, E, between an electrode sensitive to the hydrogen ion activity and a reference electrode, such as a calomel electrode or a silver chloride electrode. The combined glass electrode ideally follows the Nernst equation:

$$E = E^0 + \frac{RT}{nF} \log_e(aH);$$

$$p_H = \frac{E^0 - E}{2.303RT/F}$$

where E is a measured potential , E° is the standard electrode potential, that is, the electode potential for the standard state in which the activity is one. R is the gas constant T is the temperature in Kelvin, F is the Faraday constant and n is the number of electrons transferred, one in this instance. The electrode potential, E, is proportional to the logarithm of the hydrogen ion activity.

The operational definition of pH is officially defined by International Standard ISO 31-8 as follows: For a solution X, first measure the electromotive force E_X of the galvanic cell reference electrode | concentrated solution of KCl | | solution X | H_2 | Pt and then also measure the electromotive force E_S of a galvanic cell that differs from the above one only by the replacement of the solution X of unknown pH, pH(X), by a solution S of a known standard pH, pH(S). The pH of X is then

$$\text{pH(X)–pH(S)} = \frac{(E_s - E_x)}{2.303RT/F}$$

The difference between the pH of solution X and the pH of the standard solution depends only on the difference between two measured potentials. Thus, pH is obtained from a potential measured with an electrode calibrated against one or more pH standards; a pH meter setting is adjusted such that the meter reading for a solution of a standard is equal to the value pH(S). Values pH(S) for a range of standard solutions S, along with further details, are given in the IUPAC recommendations The standard solutions are often described as standard buffer solution. In practice it is better to use two or more standard buffers to allow for small deviations from Nernst-law ideality in real electrodes. Note that because the temperature occurs in the defining equations, the pH of a solution is temperature-dependent.

Measurement of extremely low pH values, such as some very acidic mine waters, requires special procedures. Calibration of the electrode in such cases can be done with standard solutions of concentrated sulfuric acid whose pH values can be calculated with using Pitzer parameters to calculate activity coefficients. pH is an example of an acidity function. Hydrogen ion concentrations can be measured in non-aqueous solvents, but this leads, in effect, to a different acidity function because the standard state for a non-aqueous solvent is different from the standard state for water. Superacids are a class of non-aqueous acids for which the Hammett acidity function, H0, has been developed.

p[H]

This was the original definition of Sørensen, which was superseded in favour of pH. However, it is possible to measure the concentration of hydrogen ions directly, if the electrode is calibrated in terms of hydrogen ion concentrations. One way to do this, which has been used extensively, is to titrate a solution of known concentration of a strong acid with a solution of known concentration of strong alkali in the presence of a relatively high concentration of background electrolyte. Since the concentrations of acid and alkali are known it is easy to calculate the concentration of hydrogen ions so that the measured potential can be correlated with concentrations. The calibration is usually carried out using a Gran plot. he calibration yieds a value for the standard electrode potential, E^0, and a slope factor, f, so that the Nernst equation in the form

$$E = E^0 + f\frac{RT}{nF}\log_e[H^+]$$

can be used to derive hydrogen ion concentrations from experimental measurements of E. The slope factor is usually slightly less than one. A slope factor of less than 0.95 indicates that the electrode is not functioning correctly. The presence of background electrolyte ensures that the hydrogen ion activity coefficient is effectively constant during the titration. As it is constant its value can be set to one by defining the standard state as being the solution containing the background electrolyte. Thus, the effect of using this procedure is to make activity equal to the numerical value of concentration.

The difference between p[H] and pH is quite small. It has been stated that pH = p[H] + 0.04. Unfortunately it is common practice to use the term "pH" for both types of measurement.

pOH

pOH is sometimes used as a measure of the concentration of hydroxide ions, OH-, or alkalinity. pOH is not measured independently, but is derived from pH. The concentration of hydroxide ions in water is related to the concentration of hydrogen ions by

$$[OH-] = KW / [H+]$$

where KW is the self-ionisation constant of water. Taking cologarithms

$$pOH = pKW - pH.$$

So, at room temperature pOH ˜ 14 - pH. However this relationship is not strictly valid in other circumstances, such as in measurements of soil alkalinity.

Applications

Pure water has a pH around 7; the exact values depends on the temperature. When an acid is dissolved in water the pH will be less than 7 and and when a base, or alkali is dissolved in water the pH will be greater than 7. A solution of a strong acid, such as hydrochloric acid, at concentration 1 mol dm-3 has a pH of 0. A solution of a strong alkali, such as sodium hydroxide, at concentration 1 mol dm-3 has a pH of 14. Thus, measured pH values will mostly lie in the range 0 to 14. Since pH is a logarithmic scale a difference of one pH unit is equivalent to a ten-fold difference in hydrogen ion concentration. Because the glass electrode (and other ion selective electrodes) reponds to activity, the electrode should be calibrated in a medium similar to the one being investigated.

An approximate measure of pH may be obtained by using a pH indicator. A pH indicator is a substance that changes colour around a particular pH value. It is a weak acid or weak base and the colour change occurs around 1 pH unit either side of its acid dissociation constant, or pKa, value. For example, the naturally occuring indicator litmus is red in acidic solutions (pH<7) and blue in alkaline (pH>7) solutions. Universal indicator consists of a mixture of indicators such that there is a continuous colour change from about pH 2 to pH 10. Universal indicator paper is simple paper that has been impregnated with universal indicator.

3

Marine Microbes

INTRODUCTION

The term 'Marine microbes' encompasses all microscopic organisms generally found in saltwater. Most micro-organisms are acellular and fall into the major categories of viruses, prokaryotes ('bacteria'), and protists, groups which differ considerably in biological characteristics. While representatives of these groups are found in virtually everywhere in marine waters and they play nearly every ecological role imaginable, their most important function is that they form the base of the food chain in marine ecosystems.

VIRUSES

Well-known to us as disease-causing agents, viruses are deceivingly simple organisms, little more than some nucleic acid within a protein container. They are 'parasitic particles' most about 40 nanometers in size. Viruses attach themselves to a living cell and inject a bit of nucleic acid into the cell; the injected nucleic acid directs the living cell to produce viruses. Generally, viruses are 'host-specific' only attacking or pirating a single species. They are very abundant in the sea; a tablespoon of seawater, 5 ml, commonly contains about 50 million viruses. As bacteria (or prokaryotes) are the most common potential host organisms in the sea, most viruses are bacteriophages (bacteria-consuming).

Prokaryotes

Prokaryotes are organisms without a distinct nucleus (their DNA is not bound within a membrane sac inside the cell). Typically, they are from 0.5-2 micrometers in size. Until recently known simply as 'bacteria', the 2 main groups recognized today are archaeabacteria and eubacteria which differ in the composition of their cell membranes. There appear to be no fundamental differences in the physiology or ecological roles played by the two types of prokaryotes but archaeabacteria appear do often to inhabit relatively extreme habitats such as the deep sea.

Most bacteria obtain energy by either absorbing marine dissolved organic matter through their cell membranes-osmotrophy (literally feeding through 'osmosis' in fact the material taken up is simply not obviously particulate, osmosis has little to do with the mechanisms used). However, some rely on sunlight and photosynthesis or the energy contained in some inorganic compounds (autotrophy or self-feeding). They are found in every environment, from sea ice at the poles to deep-sea hydrothermal vents. In seawater typical concentrations are about a million per ml or 5 million in a tablespoon.

Protists

Protists are eukaryotic, possessing a membrane-bound nucleus, but are single-celled or acellular organisms. The group includes all eukaryotic organisms which are not multi-cellular. Thus, it is a group of organisms united more by what they are not- multicellular- then by ancestry or common ecological characteristics. Marine protists typically range in size from 2 to 200 micrometers. Whereas viruses are parasites, and prokaryotes are osmotrophs or autotrophs, marine protist provide examples of these distinct life-styles as well as certain combinations of strategies. The different types are found in different concentrations. Protists which have chloroplasts, allowing them to perform photosynthesis thus act as autotrophs, are generally found in the highest concentrations. The larger forms (10-200 micrometers in size) include diatoms and many dinoflagellates.

The most abundant are small (1-10 micrometers long) flagellates. Autotrophic protists are restricted to the upper sunlit portion of the seas and found in abundances of a thousand per ml for the small flagellates. While they have few morphological characteristics allowing us to distinguish species, recent genetic studies suggest that small marine flagellates may be a very diverse group of organisms. Larger autrophic protists such as diatoms and dinoflagellates typically occur in concentrations of about one cell per ml.

Protists which rely on aquiring pre-formed organic matter are heterotrophic. Usually in surface waters there are about a thousand per ml of small flagellates which feed on bacteria (both autotrophic and heterotrophic prokaryotes) and 1 or 2 ciliates, oligotrichs and tintinnids) or heterotrophic dinoflagellates which feed on autotrophic protists. Besides these two large, common life-styles there are parasitic protists as well 'mixotrophic' protists. Mixotrophic protists use both photosynthesis from chloroplasts as well as feeding on pre-formed organic matter, often in the form of other protists. Some protist species retain and use the chloroplasts in the prey they eat while other protists harbor symbionts, entire autotrophic bacteria or protists.

ECOLOGICAL ROLES OF MARINE MICROBES

Primary Production

Primary production is 'first production' - the creation of organic matter. Usually it refers to the transformation, or fixation, of inorganic carbon into simple sugars using solar energy through photosynthesis. On land, the primary producers take the form of grasses, bushes, and trees and these plants are usually the most visible of all organisms within a given locality. However, in the sea (with the exception of coastal areas with seagrasses & seaweeds) the primary producers appear invisible because they microscopic and the organisms we see are high up in the food chain. The primary producers are microbes. This is not only in open water areas where the plants are in the form of the plankton (the plant plankton or phytoplankton) but it also true of many shallow areas.

Among the prokaryotes, the most abundant and important primary producers are species of the genera Synechococcus and Prochlorococcus. They are capable of reproducing once per day and form most of the 'plant' biomass in many areas of the open sea, especially in the tropics. As these cells are small (about a micrometer across) the most likely consumers of Synechococcus or Prochlorococcus are flagellate and ciliate protists.

In most coastal zones, protists form the bulk of the phytoplankton. Diatoms , dinoflagellates , and many different types of flagellates are responsible for most of the primary production. Many are large enough (> 100 micrometer) to be consumed by filter-feeding fish such as anchovies and sardines. However, throughout most of the seas, protist primary producers are small flagellates (2-10 micrometers) and fed upon by other protists, typically ciliates . The ciliate herbivores are then fed upon by larger organisms such as copepods (small crustaceans).

HERBIVORY (OR SECONDARY) AND TERTIARY PRODUCTION

Consumers of plants or primary producers, are herbivores and the production (increases in numbers or individual mass) of herbivores is called secondary production. Among marine microbes, consumers of primary producers are those which feed on autotrophic prokaryotes or autotrophic protists. Thus, the viruses which attack the autotrophic prokaryotes Synechococcus, the bacteria which absorb dissolved organic excreted by autotrophic protists such as diatoms and dinoflagellates, and the protists such as ciliates, radiolarians which feed on autotrophic protists are all consumers of primary production. However, these simple relationships exist alongside many others because among marine microbes there is not a food chain but rather a web.

Thinking in terms of food chains, more often than not marine microbes are not exactly akin to plant (phytoplankton) nor animal (zooplankton). Furthermore, relationships between different microbes are usually neither direct nor exclusive. For example, the dissolved organic matter absorbed by a bacterium is likely a mixture of that excreted by a primary producer and some from the viral lysis of another bacterium as well the excreta of yet another organism that fed on a herbivore or a primary producer. Similarly, among protists, a radiolarian may capture and ingest, more or less indifferently, a bacterium, an autotrophic flagellate, a herbivorous oligotrich ciliate, or another radiolarian . While the ecological roles are not often clear cut among marine microbes, the rest of the marine food web ultimately depends on the microbial community.

MICROORGANISM

A microorganism also spelled micro organism or micro-organism) or microbe is an organism that is microscopic (usually too small to be seen by the naked human eye). The study of microorganisms is called microbiology, a subject that began with Anton van Leeuwenhoek's discovery of microorganisms in 1675, using a microscope of his own design.

Microorganisms are incredibly diverse and include bacteria, fungi, archaea, and protists, as well as some microscopic plants and animals such as plankton, and popularly-known animals such as the planarian and the amoeba. Many scientists would not include viruses and prions, which are often classified as non-living. Most microorganisms are unicellular (or single-celled), but some multicellular organisms are microscopic, while some unicellular protists and bacteria, like *Thiomargarita namibiensis*, are macroscopic (visible to the naked eye). Microorganisms live in all parts of the biosphere where there is liquid water, including hot springs, on the ocean floor, high in the atmosphere and deep inside rocks within the Earth's crust. Microorganisms are critical to nutrient recycling in ecosystems as they act as decomposers. As some microorganisms can fix nitrogen, they are a vital part of the nitrogen cycle, and recent studies indicate that airborne microbes may play a role in precipitation and weather.

Microbes are also exploited by people in biotechnology, both in traditional food and beverage preparation, and in modern technologies based on genetic engineering. However, pathogenic microbes are harmful, since they invade and grow within other organisms, causing diseases that kill millions of people, other animals, and plants.

EVOLUTION

Single-celled microorganisms were the first forms of life to develop on earth, approximately 3–4 billion years ago. Further evolution was slow and for about three billion years in the Precambrian eon, all organisms were microscopic. So, for most of the history of life on Earth the only form of life were microorganisms Bacteria, algae and fungi have been identified in amber that is 220 million years old, which shows that the morphology of microorganisms has changed little since the triassic period.

Most microorganisms can reproduce rapidly and microbes such as bacteria can also freely exchange genes by conjugation, transformation and transduction between widely-divergent species. This horizontal gene transfer, coupled with a high mutation rate and many other means of genetic variation, allows microorganisms to swiftly evolve (via natural selection) to survive in new environments and respond to environmental stresses. This rapid evolution is important in medicine, as it has led to the recent development of 'super-bugs'-pathogenic bacteria that are resistant to modern antibiotics.

PRE-MICROBIOLOGY

The possibility that microorganisms might exist was discussed for many centuries before their actual discovery in the 17th century. The first ideas about microorganisms were those of the Roman scholar Marcus Terentius Varro in a 1st century BC book titled *On Agriculture* in which he warns against locating a homestead near swamps:

> "Because there are bred certain minute creatures which cannot be seen by the eyes, which float in the air and enter the body through the mouth and nose and there cause serious diseases".

This passage seems to indicate that the ancients were aware of the possibility that diseases could be spread by yet unseen organisms.

In *The Canon of Medicine* (1020), Abu Ali ibn Sina (Avicenna) stated that bodily secretion is contaminated by foul foreign earthly bodies before being infected. He also hypothesized that tuberculosis and other diseases might be contagious, *i.e.* that they were infectious diseases, and used quarantine to limit their spread.

When the Black Death bubonic plague reached al-Andalus in the 14th century, Ibn Khatima wrote that infectious diseases were caused by contagious "minute bodies" that enter the human body ater, in 1546, Girolamo Fracastoro proposed that epidemic diseases were caused by transferable seedlike entities that could transmit infection by direct or indirect contact, or even without contact over long distances.

All these early claims about the existence of microorganisms were speculative in nature and not based on any data or science. Microorganisms were neither proven, observed, nor correctly and accurately described until the 17th century. The reason for this was that all these early inquiries lacked the most fundamental tool in order for microbiology and bacteriology to exist as a science, and that was the microscope.

Discovery

Anton van Leeuwenhoek was the first person to observe microorganisms, using a microscope of his own design, thereby making him the first microbiologist. In doing so Leeuwenhoek would make one of the most important contributions to biology and open up the fields of microbiology and bacteriology. Prior to Leeuwenhoek's discovery of microorganisms in 1675, it had been a mystery as to why grapes could be turned into wine, milk into cheese, or why food would spoil. Leeuwenhoek did not make the connection between these processes and microorganisms, but using a microscope, he did establish that there were forms of life that were not visible to the naked eye Leeuwenhoek's discovery, along with subsequent observations by Lazzaro Spallanzani and Louis Pasteur, ended the long-held belief that life spontaneously appeared from non-living substances during the process of spoilage.

Lazzarro Spallanzani found that microorganisms could only settle in a broth if the broth was exposed to the air. He also found that boiling the broth would sterilise it and kill the microorganisms. Louis Pasteur expanded upon Spallanzani's findings by exposing boiled broths to the air, in vessels that contained a filter to prevent all particles from passing through to the growth medium, and also in vessels with no filter at all, with air being admitted via a curved tube that would not allow dust particles to come in contact with the broth. By boiling the broth beforehand, Pasteur ensured that no microorganisms survived within the broths at the beginning of his experiment. Nothing grew in the broths in the course of Pasteur's experiment. This meant that the living organisms that grew in such broths came from outside, as spores on dust, rather than spontaneously generated within the broth. Thus, Pasteur dealt the death blow to the theory of spontaneous generation and supported germ theory.

In 1876, Robert Koch established that microbes can cause disease. He did this by finding that the blood of cattle who were infected with anthrax always had large numbers of *Bacillus anthracis*. Koch also found that he could transmit anthrax from one animal to another by taking a small sample of blood from the infected animal and injecting it into a healthy one, causing the healthy animal to become sick. He also found that he could grow the bacteria in a nutrient broth, inject it into a healthy animal, and cause illness. Based upon these experiments, he devised criteria for establishing a causal link between a microbe and a disease in what are now known as Koch's postulates. Though these postulates cannot be applied in all cases, they do retain historical importance in the development of scientific thought and can still be used today.

Classification and Structure

Microorganisms can be found almost anywhere in the taxonomic organisation of life on the planet. Bacteria and archaea are almost always microscopic, while a number of eukaryotes are also microscopic, including most protists, some fungi, as well as some animals and plants. Viruses are generally regarded as not living and therefore are not microbes, although the field of microbiology also encompasses the study of viruses.

Prokaryotes

Prokaryotes are organisms that lack a cell nucleus and the other organelles found in eukaryotes. Prokaryotes are almost always unicellular, although some species such as myxobacteria can aggregate into complex structures as part of their life cycle. These organisms are divided into two groups, the archaea and the bacteria.

Bacteria

Bacteria are the most diverse and abundant group of organisms on Earth. Bacteria inhabit practically all environments where some liquid water is available and the temperature is below +140 °C. They are found in sea water, soil, air, animals' gastrointestinal tracts, hot springs and even deep beneath the Earth's crust in rocks. Practically all surfaces which have not been specially sterilized are covered in bacteria. The number of bacteria in the world is estimated to be around five million trillion trillion, or 5 × 1030.

Bacteria are practically all invisible to the naked eye, with a few extremely rare exceptions, such as *Thiomargarita namibiensis*. They are unicellular organisms and lack membrane-bound organelles. Their genome is usually a single loop of DNA, although they can also harbor small pieces of DNA called plasmids. These plasmids can be transferred between cells through bacterial conjugation. Bacteria are surrounded by a cell wall, which provides strength and rigidity to their cells. They reproduce by binary fission or sometimes by budding, but do not undergo sexual reproduction. Some species form extraordinarily resilient spores, but for bacteria this is a mechanism for survival, not reproduction. Under optimal conditions bacteria can grow extremely rapidly and can double as quickly as every 10 minutes

Archaea

Archaea are also single-celled organisms that lack nuclei. In the past, the differences between bacteria and archaea were not recognised and archaea were classified with bacteria as part of the kingdom Monera. However, in 1990 the microbiologist Carl Woese proposed the three-domain system that divided living things into bacteria, archaea and eukaryotes. Archaea differ from bacteria in both their genetics and biochemistry. For example, while bacterial cell membranes are made from phosphoglycerides with ester bonds, archaean membranes are made of ether lipids.

Archaea were originally described in extreme environments, such as hot springs, but have since been found in all types of habitats Only now are scientists beginning to appreciate how common archaea are in the environment, with crenarchaeota being the most common form of life in the ocean, dominating ecosystems below 150 m in depth These organisms are also common in soil and play a vital role in ammonia oxidation.

Eukaryotes

All living things which are *individually* visible to the naked eye are eukaryotes (with few exceptions, such as *Thiomargarita namibiensis*), including humans. However, a large number of eukaryotes are also microorganisms. Unlike bacteria and archaea, eukaryotes contain organelles such as the cell nucleus, the Golgi apparatus and mitochondria in their cells. The nucleus is an organelle which houses the DNA that makes up a cell's genome. DNA itself is arranged in complex chromosomes. Mitochondria are organelles vital in metabolism as they are the site of the citric acid cycle and oxidative phosphorylation. They evolved from symbiotic bacteria and retain a remnant genome Like bacteria, plant cells have cell walls, and contain organelles such as chloroplasts in addition to the organelles in other eukaryotes. Chloroplasts produce energy from light by photosynthesis, and were also originally symbiotic bacteria.

Unicellular eukaryotes are those eukaryotic organisms that consist of a single cell throughout their life cycle. This qualification is significant since most multicellular eukaryotes consist of a single cell called a zygote at the beginning of their life cycles. Microbial eukaryotes can be either haploid or diploid, and some organisms have multiple cell nuclei (see coenocyte). However, not all microorganisms are unicellular as some microscopic eukaryotes are made from multiple cells.

Protists

Of eukaryotic groups, the protists are most commonly unicellular and microscopic. This is a highly diverse group of organisms that are not easy to classify Several algae species are multicellular protists, and slime molds have unique life cycles that involve switching between unicellular, colonial, and multicellular forms he number of species of protozoa is uncertain, since we may have identified only a small proportion of the diversity in this group of organisms.

All animals are multicellular, but some are too small to be seen by the naked eye. Microscopic arthropods include dust mites and spider mites. Microscopic crustaceans include copepods and the cladocera, while many nematodes are too small to be seen with the naked eye. Another particularly common group of microscopic animals are the rotifers, which are filter feeders that are usually found in fresh water. Micro-animals reproduce both sexually and asexually and may reach new habitats as eggs that survive harsh environments that would kill the adult animal. However, some simple animals, such as rotifers and nematodes, can dry out completely and remain dormant for long periods of time.

Fungi

The fungi have several unicellular species, such as baker's yeast (Saccharomyces cerevisiae) and fission yeast (Schizosaccharomyces pombe). Some fungi, such as the pathogenic yeast Candida albicans, can undergo phenotypic switching and grow as single cells in some environments, and filamentous hyphae in others. Fungi reproduce both asexually, by budding or binary fission, as well by producing spores, which are called conidia when produced asexually, or basidiospores when produced sexually.

Plants

The green algae are a large group of photosynthetic eukaryotes that include many microscopic organisms. Although some green algae are classified as protists, others such as charophyta are classified with embryophyte plants, which are the most familiar group of land plants. Algae can grow as single cells, or in long chains of cells. The green algae include unicellular and colonial flagellates, usually but not always with two flagella per cell, as well as various colonial, coccoid, and filamentous forms. In the Charales, which are the algae most closely related to higher plants, cells differentiate into several distinct tissues within the organism. There are about 6000 species of green algae.

Habitats and Ecology

Microorganisms are found in almost every habitat present in nature. Even in hostile environments such as the poles, deserts, geysers, rocks, and the deep sea, some types of microorganisms have adapted to the extreme conditions and sustained colonies; these organisms are known as extremophiles. Extremophiles have been isolated from rocks as much as 7 kilometres below the earth's surface and it has been suggested that the amount of living organisms below the earth's surface may be comparable with the amount of life on or above the surface Extremophiles have been known to survive for a prolonged time in a vacuum, and can be highly resistant to radiation, which may even allow them to survive in space. Many types of microorganisms have intimate symbiotic relationships with other larger organisms; some of which are mutually beneficial (mutualism), while others can be damaging to the host organism (parasitism). If microorganisms can cause disease in a host they are known as pathogens.

Extremophiles

Extremophiles are microorganisms which have adapted so that they can survive and even thrive in conditions that are normally fatal to most lifeforms. For example, some species have been found in the following extreme environments:

- Temperature: as high as 130 °C (266 °F) as low as -17 °C (1.4 °F)
- Acidity/alkalinity: less than pH 0 up to pH 11.5
- Salinity: up to saturation
- Pressure: up to 1,000-2,000 atm, down to 0 atm (e.g. vacuum of space)

Radiation: up to 5kGy

Extremophiles are significant in different ways. They extend terrestrial life into much of the Earth's hydrosphere, crust and atmosphere, their specific evolutionary adaptation mechanisms to their extreme environment can be exploited in bio-technology, and their very existence under such extreme conditions increases the potential for extraterrestrial life.

Extremophiles

- Notable extremophiles Chloroflexus aurantiacus, Deinococcus radiodurans, Deinococcus-Thermus, Paralvinella sulfincola, Pompeii worm, Pyrococcus furiosus Snottite, Strain 121, Thermus aquaticus, Thermus thermophilus, Spirochaeta americana, Tardigrada.
- Archaeoglobaceae, Berkeley Pit, Crenarchaeota, Grylloblattidae, Halobacteria.
- Halobacterium, Hydrothermal vent, Methanopyrus, Radioresistance, Thermostability, Thermotogae.

Soil Microbes

The nitrogen cycle in soils depends on the fixation of atmospheric nitrogen. One way this can occur is in the nodules in the roots of legumes that contain symbiotic bacteria of the genera Rhizobium, Mesorhizobium, Sinorhizobium, Bradyrhizobium, and Azorhizobium.

Symbiotic Microbes

Importance

Microorganisms are vital to humans and the environment, as they participate in the Earth's element cycles such as the carbon cycle and nitrogen cycle, as well as fulfilling other vital roles in virtually all ecosystems, such as recycling other organisms' dead remains and waste products through decomposition. Microbes also have an important place in most higher-order multicellular organisms as symbionts. Many blame the failure of Biosphere 2 on an improper balance of microbes.

Use in Food

Microorganisms are used in brewing, winemaking, baking, pickling and other food-making processes. They are also used to control the fermentation process in the production of cultured dairy products such as yogurt and cheese. The cultures also provide flavour and aroma, and inhibit undesirable organisms.

Use in Water Treatment

Specially-cultured microbes are used in the biological treatment of sewage and industrial waste effluent, a process known as bioaugmentation.

Use in Energy

Microbes are used in fermentation to produce ethanol, and in biogas reactors to produce methane. Scientists are researching the use of algae to produce liquid fuels and bacteria to convert various forms of agricultural and urban waste into usable fuels.

Use in Science

Microbes are also essential tools in biotechnology, biochemistry, genetics, and molecular biology. The yeasts (Saccharomyces cerevisiae) and fission yeast (Schizosaccharomyces pombe) are important model organisms in science, since they are simple eukaryotes that can be grown rapidly in large numbers and are easily manipulated. They are particularly valuable in genetics, genomics and proteomics Microbes can be harnessed for uses such as creating steroids and treating skin diseases. Scientists are also considering using microbes for living fuel cells, and as a solution for pollution.

Biological Warfare

In the Middle Ages, diseased corpses were thrown into castles during sieges using catapults or other siege engines. Individuals near the corpses were exposed to the deadly pathogen and were likely to spread that pathogen to others.

Importance in Human Health

Microorganisms can form an endosymbiotic relationship with other, larger organisms. For example, the bacteria that live within the human digestive system contribute to gut immunity, synthesise vitamins such as folic acid and biotin, and ferment complex indigestible carbohydrates.

Diseases and Immunology

Pathogenic Microbes

Microorganisms are the cause of many infectious diseases. The organisms involved include pathogenic bacteria, causing diseases such as plague, tuberculosis and anthrax; protozoa, causing diseases such as malaria, sleeping sickness and toxoplasmosis; and also fungi causing diseases such as ringworm, candidiasis or histoplasmosis. However, other diseases such as influenza, yellow fever or AIDS are caused by pathogenic viruses, which are not usually classified as living organisms and are not therefore microorganisms by the strict definition. As of 2007, no clear examples of archaean pathogens are known although a relationship has been proposed between the presence of some methanogens and human periodontal disease.

Importance in Ecology

Decomposition Microbes are critical to the processes of decomposition required to cycle nitrogen and other elements back to the atural world.

Hygiene

Hygiene is the avoidance of infection or food spoiling by eliminating microorganisms from the surroundings. As microorganisms, particularly bacteria, are found practically everywhere, this means in most cases the reduction of harmful microorganisms to acceptable levels. However, in some cases it is required that an object or substance be completely sterile, i.e. devoid of all living entities and viruses. A good example of this is a hypodermic needle. In food preparation microorganisms are reduced by preservation methods (such as the addition of vinegar), clean utensils used in preparation, short storage periods or by cool temperatures. If complete sterility is needed, the two most common methods are irradiation and the use of an autoclave, which resembles a pressure cooker. There are several methods for investigating the level of hygiene in a sample of food, drinking water, equipment etc. Water samples can be filtrated through an extremely fine filter. This filter is then placed in a nutrient medium. Microorganisms on the filter then grow to form a visible colony.

Harmful microorganisms can be detected in food by placing a sample in a nutrient broth designed to enrich the organisms in question. Various methods, such as selective media or PCR, can then be used for detection. The hygiene of hard surfaces, such as cooking pots, can be tested by touching them with a solid piece of nutrient medium and then allowing the microorganisms to grow on it. There are no conditions where all microorganisms would grow, and therefore often several different methods are needed. For example, a food sample might be analyzed on three different nutrient mediums designed to indicate the presence of "total" bacteria (conditions where many, but not all, bacteria grow), molds (conditions where the growth of bacteria is prevented by e.g. antibiotics) and coliform bacteria (these indicate a sewage contamination).

In Fiction

Microorganisms have frequently played an important part in science fiction, both as agents of disease, and as entities in their own right.

Some notable uses of microorganisms in fiction include:

- The War of the Worlds, where microorganisms play important thematic and plot-related roles. Fantastic Voyage, in which some scientists are miniaturised to microscopic size and observe micro-organisms from a new perspective Blood Music, in which a colony of microorganisms is given intelligence.
- The Andromeda Strain, in which extraterrestrial microorganisms kill several people The White Plague, iscreated and released in vengeance by John Roe O'Neill for the death of his wife and children, it is designed to kill only women.

- Twelve Monkeys, James Cole (Bruce Willis) searches for a pure germ in the past, which creates a deadly plague in the future. Also, Brad Pitt (as Jeffery Goines) discusses his germaphobia.

Microbial Food Webs

Importance

Life in the ocean is microbe based. A large fraction of the biomass and biological activity in the ocean is microscopic, and it is an important part of earth's carbon and nitrogen cycles. There are 100 million times more bacteria in the ocean than stars in the known universe, and there are a thousand times more viruses than bacteria. Yet we know little about the relationships between microbes and their environment because only one tenth of one percent of the bacteria have ever been cultured. The traditional marine food web exists in a sea of microbes and dissolved carbon-based matter.

The entire microbial food web, including protozoan microzooplankton, is typically some five to ten times the mass of all multicellular marine organisms. The potential metabolic dominance of microorganisms is even greater than their biomass suggests. Put in more understandable terms, a mass of B. *natrigens* [a bacteria] equal to 100 humans would have an enegy throughput of about a gigawatt, much the same as a nuclear power station. This metabolic potential under optimal circumstances would be rarely, if ever, achieved in nature for a number of reasons, notably the low concentrations of organic nutrients.

One approach to understanding the relationships between micro-organisms and their ocean environment is to examine their DNA using, for example, the shotgun technique. In 2003, Craig Venter and colleagues used the technique to analyze the DNA sequences of micro-organisms in water collected from the Sargasso Sea near Bermuda and found.

1800 genomic species based on sequence relatedness, including 148 previously unknown bacterial phylotypes. We have identified over 1.2 million previously unknown genes represented in these samples, including more than 782 new rhodopsin-like photoreceptors.

Another approach has been to measure the number of microbes, their productivity, and the amount of particulate matter in the water to calculate how much the microbes contribute to various parts of the oceanic food web. This approach shows that microbes, especially the picoplankton, microbial plankton with sizes ranging from 0.2 to 2.0 microns, are as important as the traditional marine food web.

The microbial food web (loop on the left) is an important part of the marine food web. Picoplankton feed larger zooplankton directly (path 6) and indirectly (paths 3, 4, and 6), and they contribute to particulate and dissolved carbon in the ocean.

The staggering variety of microbes discovered through shotgun DNA analysis shows how little we know about the microbial web. Craig Venter, in a second voyage of discovery from New England to the southwest Pacific, returned with billions of bases of DNA sequences.

Venter's team has found evidence of so many new microbial species that the researchers want to redraw the tree of microbial life. [They] identified more than 400 microbial species new to science. The vast majority of the microbes that found themselves snared in Venter's filters were genetically unique. Based on their best guess as to the beginning and end of each gene teased out from the DNA sequences, Venter Institute computational biologist Shibu Yooseph and his colleagues have concluded that the DNA encodes 6.12 million hypothetical proteins. That finding almost doubles the number of known proteins in a single stroke. Most of the predicted proteins are of unknown function, and a quarter of them have no similarity to any known proteins. By comparing predicted amino acid sequences with those of known proteins, they found a surprising abundance of signaling proteins thought to be used only by multicellular organisms. Among the hypothetical proteins from their marine samples, the researchers found 28,000 of the so-called eukaryotic protein kinases, as well as another 19,000 of a group that are highly similar to these kinases— triple the number previously known.

One 10-month expedition that sampled the ocean at just 41 locations returned with hundreds of new species able to make millions of proteins that were previously unknown, including tens of thousands that were thought to occur only in higher, multicelled organisms. And this is from only a tiny part of the ocean. Clearly, the microbial web is almost unknown. A vast new frontier of life awaits discovery.

The microbial web includes bacteria and archaea (organisms without a cell nucleus) and very small eukaryotes (organisms having cells with a nucleus), ranging in size from 0.01 micrometers for viruses to 20 micrometers for ciliates, living in a dilute broth of carbon-based matter ranging from individual molecules to tangled polymers to colloids to clumps of dead matter called marine snow.

The microbial loop: impressionist version. A bacteria-eye view of the ocean's sunlit layer. Seawater is an carbon-based matter continuum, a gel of tangled polymers with embedded strings, sheets, and bundles of fibrils and particles, including living organisms, as "hotspots." Bacteria (red) acting on marine snow (black) or algae (green) can control sedimentation and primary productivity; diverse microniches (hotspots) can support high bacterial diversity.

Viruses

Most of the biomass in the ocean is made up of viruses. They attack and kill bacteria, archaea and other micro-organisms, which releases the dead's cell contents, adding to the carbon-based material in the water.

Patrick Forterre, an evolutionary biologist at the University of Paris-Sud in Orsay, France, believes that viruses are at the very heart of evolution. Viruses, Forterre argues, bequeathed DNA to all living things. Trace the ancestry of your genes back far enough, in other words, and you bump into a virus.

Viruses can be considered to be on the threshold between living and non-living. In a sense, they can cause themselves to be reproduced, so they could be considered as being "alive". But they can only do this by using the machinery of a host cell. Without a host they exist only as inert, almost crystal-like, "non living" particles in the environment. They range generally from 17 to 1000 nm in size (a nanometer = 10-9m). Viruses are generally host specific, that is a given virus will generally affect only a certain species or group of related species, or even only certain kinds of cells within a species.

Bacteria

The ocean is filled with a countless mob of bacteria with a staggering variety of genes according to Bohannon (2007). The bacteria are a critical part of the marine food web, processing more than half of all the flow of carbon-based matter. Some use sunlight for energy, the photosynthetic bacteria, some use both sunlight and carbon-based material for energy, the mixotrophs, and some attack other organisms, the heterotrophs. See the vocabulary section below.

It [was] generally thought that pelagic bacteria passively receive dissolved carbon-based matter leaking from the grazing food chain and diffused homogeneously. However, recent studies on the complex nature of the carbon-based matter field and behavioural strategies of bacteria have suggested that bacteria do not use only preformed dissolved carbon-based matter; they also attack all carbon-based matter, even live organisms, thus liberating dissolved carbon-based matter. Hence, bacterial attack can profoundly modify the biogeochemical behaviour of carbon-based matter in ways not inferred from measuring cumulative carbon-based matter fluxes into bacteria.

The cyanobacteria are the most common type of bacteria in the ocean, and SAR11 (Pelagibacter clade of the Synechoccus types of bacteria) is the most common organism on earth, accounting for roughly 25% of all the bacteria in the ocean. Prochlorococcus and Synechoccus species dominate the microbial ecology of the ocean. Yet, they were mostly unknown until Sallie W. Chisholm, a professor at the Massachusetts Institute of Technology, first discovered Prochloroccus in 1988.

The marine bacterium SAR11, the most common organism on earth. It was discovered by Stephen Giovannoni of Oregon State University in 2002.

Vibrio Vulnificus

Vibrio vulnificus in coastal waters, including Texas lagoons, causes a flesh-eating disease (necrotizing fasciitis) that can kill in a few days without prompt treatment. If you have a cut that becomes infected after you were in warm lagoon water, go directly to an emergency room, especially if the infection spreads. Some infections can increase from the size of a nickel to the size of a quarter in a few hours.

V. vulnificus bacterial infection that developed one day after a fish bone injury on the fourth finger of the left hand (arrow) in a 45-year-old patient with uremia. From Bacteriology at UW-Madison. A Danish website on Vibrio vulnificus has more images.

Archaea

They look very much like bacteria, but they are no more closely related to bacteria than we are to bacteria. The Archaea dominate some oceanic areas, especially below the sunlit zones, where "they make up typically 40% of deep-sea organisms and the deep sea is by far the largest habitat on earth." – Jed Fuhrman, University of Southern California, quoted in Copley (2002). Overall, "20% of all the picoplankton cells in the world ocean appear to be represented by one specific clade, the pelagic [open-ocean] crenarchaeota".

They also dominate subsea sediments:

The deep biosphere is large, and, at least in aggregate, it is 'alive'. The data almost defy imagination — on the order of a million cells in every cubic centimetre of sediment buried half a kilometre below the sea floor, or more than 50% of all microbial cells on Earth. The overall order of magnitude is the same as the biomass of all surface plant life. In investigating what kind of cells they are, Lipp et al. (2008) conclude that, deeper than 1 metre below the sea floor, it is Archaea, not Bacteria, that contribute the bulk of sedimentary microbial biomass. If true, Archaea would be the most abundant cell type in the marine system, rivalling Bacteria in total numbers globally.

Transmission electron micrograph of *Nitrosopumilus maritimus*, (also known as SCM-1), an archae of the kingdom crenarchaeota. Very similar, open-ocean crenarchaeota are thought to have an important influence on the nitrogen cycle. Scale bar represents 0.1 micrometre.

Ciliates

They are microplankton of the subclass Choreotricha that are filter feeders. A paramecium is a common example of a cilate. A marine ciliate of the genus strombidium.From Gulf of Alaska Plankton Photos.

Some Vocabulary Organisms can be classified by how they obtain energy and nutrients.

- Autotrophs synthesize their own nutrients. Phototrophs such as phytoplankton use sunlight and inorganic nutrients to synthesize carbon-based nutrients.
- Chemotrophs get energy from inorganic chemical reactions, such as the oxidation of methane (methanotrophs). Mixotrophs are organisms that can both photosynthesize (phototrophic) and ingest particulate carbon-based matter and bacteria phagotrophic).
- Heterotrophs are organisms that must eat other organisms to obtain nutrients. People are heterotrophs. Bacteriovores eat bacteria. Herbivores eat plants.
- Carnivores eat animals.

Microbiology—The Long Version

The range of environmental conditions that support life on Earth have been considerably expanded in recent decades. Microbial life is thought to exist in nearly all earthly near-surface habitats below a temperature of 150° C. Life has been detected in a wide variety of environments from hot vents to very cold and icy environments. Modern phylogenetic information suggests that the "Tree of Life" includes bacterial (formerly "eubacterial"), archaeal (formerly "archaebacterial") and eukaryal (formerly "eukaryotic") domains. It is believed that early phylogenetic separation and evolution produced a complex history of life's evolution. Organisms that live in extreme environments are often located near the base of the "Tree of Life" representing some of its earliest and possibly most primitive forms. The exact relationship of the major domains is unresolved, the monophyly of Archaea is uncertain, and recent evidence for ancient lateral transfers of genes concerned with metabolism indicates that the web of life is a complex network of phylogenetic relationships. Investigation of organisms from extreme environments provide important information for understanding how life arose, developed, and evolved.

The environmental boundaries of habitats that support life on Earth can be viewed in a multi-dimensional space with properties such as temperature, pressure, inorganic and organic nutrients, and energy fluxes as variables. The temperature and pressure limits to life are now well constrained. For temperature, hyper-thermophilic prokaryotic archaea, with measurable growth rates at 121°C, have been cultured from deep-sea hydrothermal systems. There is evidence for biological activity at even higher temperatures. At the lower end of the temperature scale, microbial activity has been detected in hyper-saline sea ice at -13°C and may be viable to even colder temperatures. Within the limits of accessible habitats, pressure by itself does not appear to preclude microbial growth. Certain bacteria require elevated hydrostatic pressure and cannot grow at one atmosphere pressure. Less well studied and understood are the limits of life with respect to water activity, nutrient or energy limitations, exposure to strongly ionizing radiation, and aquatic habitats where gas is present as gas hydrate. Available energy places an ultimate constraint on microbial processes in otherwise "life-permissive" habitats and it is for this reason that subglacial environments are unique "natural laboratories." The darkness and isolation of subglacial environments add constraints on energy and resources that likely limit primary productivity more than in other habitats. As in the deep earth biosphere, geothermal heat and chemical fluxes, if they are present, will be of profound importance for subglacial microbiology. Subglacial lakes are thought to be relatively stable environments that may contain ancient ecosystems. As with surface lakes, the physical and hydrological stability of lakes over time will be a major determinant of the kind of ecosystems present.

Various scenarios have been proposed to describe how life might exist in subglacial environments. In one scenario, microbial life is solely dependent on the reservoir of reduced carbon present at the time of "closure." For example, if Lake Vostok were a typical alpine lake, the original energy stored in the lake would sustain microbial life for only a limited amount of time. Organic carbon reserves within the underlying sediments could supplement the pelagic food web via upward diffusion. In this scenario, Lake Vostok would be one of the few habitats on Earth where reduced energy flux precludes a functional ecosystem. The geochemical disequilibria that characterize biotic-controlled habitats would approach thermodynamic equilibrium, a truly unique condition on the Earth but one hypothesized for the ice-covered "ocean" of the Jovian moon Europa. The ultimate outcome would be the collapse of the ecosystem.

In another scenario, refractory organic matter in the pre-glaciation ecosystem serves as an "energy buffer" to support reduced life process rates. Over time, the system would spiral down in terms of potential energy, altering the ecological structure and functioning of microbial communities. The present day conditions for life would be the most ultra-oligotrophic known on Earth. Very low standing stocks of microorganisms would be present and the resident communities would exhibit low rates of energy transformation, growth, and reproduction. In this scenario, microorganisms may express a "starvation response" that has been observed in the laboratory when marine bacteria experience extreme energy limitations. The community selected for would be distinct from the pre-glaciation environment favoring small, slow growing (or non-growing, spore forming), low-maintenance prokaryotes. As organism density dropped below a critical threshold of approximately 1000 cells per millilitre, cell-to-cell interactions, including key ecological processes such as quorum sensing, syntrophy, grazing, symbiosis, and viral infections would cease to exist. New ecological theory would be required to describe ecosystem processes under these conditions. These two low energy scenarios may not be operative in some subglacial environments if the tectonic framework is active in terms of material and energy transfer. Tectonically triggered fluxes of crustal fluids might lead to localized environments similar to those surrounding deep-sea hydrothermal vents and oil and gas seeps in the Gulf of Mexico. The structure and function of microbial communities that might evolve under this scenario will depend on the composition, the frequency, and the amount of heat and chemicals infused into the system.

The concentration of dissolved gases in lake water will exert an important control on microbiological processes in subglacial environments. High oxygen concentrations have profound effects on microorganisms. One possible response is that microorganisms adapt by producing high concentrations of enzymes to deal with excessive oxygen radicals; e.g., peroxidases, catalases, and superoxide dismutases. Although lake water oxygen concentration estimates assumed no removal processes within the lake, it has been hypothesized that aerobic conditions in the upper portion of the water column could support autotrophic and heterotrophic metabolic lifestyles. Deeper portions of the lake and sediment are likely to be anoxic if oxygen removal processes are operative in the overlying water or surficial sediments. For Lake Vostok, microbiological metabolism may be heterotrophic (based on organic carbon) or chemolithoautrophic (based on CO_2 and chemical energy). Accretion ice DNA sequences imply that organisms capable of both types of metabolism may be present in Lake Vostok. Controversy remains as to the concentration of dissolved organic carbon that may exist within the water column and its potential to fuel heterotrophic metabolism.

If surface lakes lend clues to the character of subglacial lakes, subglacial sediments will offer charged surfaces for cell attachment and chemical concentration, provide organic and nutrient-rich conditions and exhibit a range of reduction-oxidation conditions. In turn, sedimentary habitats support a multitude of metabolic life styles. The physical/chemical gradients encountered in many sediments result in biomass and activity rates that are almost invariably higher than those in the overlying water column. Biogeochemical transformations produce chemical gradients within the sediment that affect the properties of the overlying water column. For example, organisms utilizing oxygen as a terminal electron acceptor can deplete the local environment leading to sub-oxic or anoxic conditions. Sediment biogeochemistry also produces chemicals (e.g., CH_4, H_2S, NH_{4+}, PO_{4-3}) that diffuse into the water column supplying electron donors and nutrients to water column bacteria. As with surface lakes, sediments are expected to be an important component of some subglacial environments (not all subglacial environments may contain sediments). Often unique microbiological communities reside at anoxic/oxic interfaces within sediments taking advantage of both environments to meet their energy and nutritional needs.

Subglacial lake environments are expected to be some of the, if not the, lowest energy and nutrient environments on Earth. Survival in such environments constitutes a significant challenge for life and presents unique opportunities to study life at its limits.

While results from analogous setting and accreted ice studies convince many that it is more likely that microbial life is present in Lake Vostok than not, this first order question must be resolved early in any exploration programme before further more detailed microbiological studies are undertaken. The detection of life in possibly the lowest biomass environments on Earth is a substantial technological challenge and protocols must be developed to detect life in very dilute environments. These protocols should include proven microbiological and biochemical methods. Other supportive evidence will be provided by the limnology and biogeochemical studies described elsewhere in this report as biological activity causes physical and chemical profiles to depart from predicted equilibrium values. If life is unambiguously detected, it will then be important to establish the amount (biomass), distribution, and diversity of organisms in subglacial lake environments. This will require "clean" sampling of water columns, sediments, and ice. Retrieval and analysis of samples at *in situ* conditions will present special technological challenges as well.

There is considerable evidence from both permanently ice-covered systems such as the McMurdo Dry Valley lakes and the seasonally ice-covered maritime Antarctic lakes of Signy Island that community structure differs substantially between lakes even for lakes in close geographical proximity. These differences are driven by the physical and geochemical conditions associated with each lake. Once life is identified in subglacial environments, the diversity of microorganisms within and between lakes will be a subject of interest. The metabolism and physiological capability of subglacial microorganisms will lend clues to special adaptive strategies for surviving in these harsh conditions.

The study of life in subglacial environments will contribute new information on life at the edges of survivability, furthering our understanding of the diversity of life on Earth, its evolution, and its response to environmental processes over geological time. Life in extreme environments results in the expression of unique adaptations and the studies proposed here will contribute further knowledge about how adaptive strategies allow for not only survival but growth under some of the most difficult environmental conditions on Earth. Subglacial environments are also analogues whose study will inform the search for life on other icy planetary bodies in our solar system.

4

Marine Viruses

INTRODUCTION

Viruses are parasites of living cells which invade cells and then use their biological machinery to propagate. Most viruses in the ocean consist of nucleic acids surrounded by a protein coat (called a capsid). Viruses are thought to have appeared several times during the evolution of cellular organisms as suggested by the type (DNA vs RNA) or structure of the nucleic acid (double-strand vs single strand; sense vs antisense). In marine environments, they are more numerous than the most abundant cell types, the prokaryotes. Prokaryotes are single-celled organisms without a nucleus and are now considered to consist of the two so-called domains of life, Bacteria and Archaea. Viruses are a major factor for the mortality of marine prokaryotes.

Most viruses in the ocean are thought to infect prokaryotes. They are typically DNA-containing viruses and, if they infect bacteria, are called bacteriophages or "phages" for short. Most phages, at least in the surface ocean, belong to three phage groups of the tailed phages. Some of these phages look like moon landers. The morphological diversity of the phages is high: they vary in capsid size as well as tail length and structure and can have appendices. In the deep ocean around hydrothermal vents spindle forms are found, which are typical for some groups of viruses infecting Archaea. In surface waters, very large viruses can be found, often with capsid diameters of 200-400 nm (0.000002-0.000004 mm). These viruses are larger than some prokaryotic cells. They probably infect single-celled photosynthetic organisms (phytoplankton) of the third domain of life, the Eukarya (cells with a nucleus). Even larger virus-like particles (700 nm) have been found in the food vacuoles of radiolarians (skeleton-forming protists). Viruses also infect metazoans and benthic plants. For example, some epidemic diseases of fish and seals have been caused by viruses. It is likely that viruses infect all cellular organisms in the ocean, from prokaryotes to whales.

THE GENETIC DIVERSITY OF VIRUSES

Only a tiny portion of the viruses from the ocean have been characterized. One of the reasons for that is that it was only 15-20 years ago that scientists realized that viruses play a key role in the functioning of the ocean. Another reason is that for a characterization of viruses, their host or host cells have to be available. This is a problem, since most prokaryotes have not been cultivated in the controlled environment of a laboratory. However, more emphasis is currently put into isolation of so called virus-host systems.

To circumvent the isolation and cultivation problems, cultivation-independent methods have been developed. These are based on the detection of specific genes that are found exclusively in viruses. A major focus in this area has been phages infecting a group of photosynthesizing bacteria, the cyanobacteria (formerly known as blue-green algae). Cyanobacteria can contribute significantly to photosynthetic carbon fixation (primary production). Using sequence analysis, many new types of these phages (called cyanophages) have been detected. Some of these sequences seem to be specific for environments such as estuaries or the open ocean. Also, a specific depth distribution in the water column is found. For example, specific sequences of cyanophages could only be detected in the deep chlorophyll maximum (DCM), a Chlorophyll-rich layer closely associated to the thermocline (ca 75 to 200 m depth). Many new sequences from uncultivated algal viruses (Phycodnaviridae) could be found as well. Sequence analysis and isolation of viruses showed that some viruses have a cosmopolitan distribution, whereas others seem have a more restricted geographical range.

Recently, attempts have been made to sequence all the genes from all of the viruses from environmental samples. Using this community or metagenomic approach, between 370 and 7100 viral types have been detected in two coastal samples. These DNA viruses mainly infected prokaryotes. RNA virus metagenomics showed there is also a large, previously unknown diversity of RNA viruses in the coastal ocean. The RNA viruses most likely infect eukaryoytes. Metagenomics suggest that viruses represent the largest unknown sequence space in the ocean. It has also been suggested based on metagenomics data that the coastal marine sediments are the most diverse biological system characterized so far. This means that for the vast majority of the sequences, it is not known what they are encoding for.

The studies have revealed a surprising functional diversity of viruses and the presence of evolution of specific viral genes. For example, viruses have genes to repair DNA damage such as caused by sunlight. Homologs of these repair genes were found in very diverse viruses such as the T4 phage infecting Escherichia coli and an algal virus infecting the symbiotic algae Chlorella. However, this gene has not been found in the genome of a cellular organisms. Cyanophage can also carry genes involved in the process of photosynthesis (which is normally attributed to only algae and plants). This is surprising, since viruses have no metabolism of their own. From observations in lab studies, it appears that these cyanophage-encoded photosynthesis genes force the infected host cell to continue with photosynthesis until shortly before cell lysis. This probably increases the energy available within the infected cell so that a maximum of new cyanophage particles can be produced. Thus, viruses do interfere with host functions in a much more sophisticated way than previously thought.

VIRUS LIFE CYCLES

Viruses have several different life (or replication) cycles. The two most dominant are the lytic and the lysogenic (latent) cycle. Lytic (or virulent) viruses infect a cell (or tissue), replicate (i.e. make new viruses) and are released by lysis the host cell. Temperate viruses infect the host and the viral DNA stays within the host cell, often in the host genome. It is said that they lysogenize the host. For eukaryotes typically the term latency is used. The viral genome replicates along with the host genomes, until some factors, such as DNA damage, induce the lytic cycle. For prokaryotes in the ocean, it has been suggested that in nutrient-rich waters (characterized by a high abundance of hosts) lytic phages dominate, whereas in nutrient-poor waters (characterized by a low abundance of hosts), lysogeny dominates.

EFFECT OF VIRUSES ON HOST DIVERSITY

It is well established, albeit not fully appreciated, that the activity of phages has a very strong influence on the diversity and distribution of bacteria. Viruses play a role in the evolution of the host by virus-mediated gene transfer to hosts. However, they also play a role in the maintenance of host diversity at ecological scales. Experimental data have shown that lytic viruses can affect the diversity of host communities of Bacteria and Archaea. Conceptual and mathematical models suggest that viruses can influence the winner in competitions for nutrients. In this scenario, once a microbe becomes dominant, the increase in abundance increases its contacts with viruses leading to significant increases in infection and subsequent lysis, which then control its abundance. This model has been called the 'kill the winner' hypothesis. As a consequence, viruses may sustain a high population diversity amongst their hosts by controlling the growth of organisms that would extensively proliferate due to advantages in nutrient acquisition. This would allow for the survival of less competitive but virus-resistant microbes.

EFFECT OF VIRUS ON ENVIRONMENTAL CHEMISTRY

Viruses in marine environments also play many important roles in the ecology of these ecosystems. Perhaps one of their most important roles is in system biogeochemistry, as through the process of infecting and lysing their hosts, viruses convert both macronutrients (nitrogen and phosphorus) as well as micronutrients (e.g., iron) from particulate forms to dissolved forms that are then available for actively growing phytoplankton and bacteria to consume. Through this process viruses work not only to recycle these nutritional elements, but also to return these elements to the water column in chemical compounds that may not be typically produced by other processes. When bacteria consume organic material, they often remineralise nutrients as inorganic forms back into the water column. However, when viruses lyse bacteria, the nutrients are released in the water column within the organic compounds which were present in the cell. These organically-complexed nutrients released to the marine community during virus lysis therefore need to be assimilated into cells by different biochemical pathways than those released during the remineralization of organic material. As such this process may alter the selective pressures on the marine microbial community in terms of having the necessary pathways to assimilate the chemical forms these obligate nutrients. Such a change in selective pressure will also affect the community composition of the plankton.

5

Marine Microorganisms

INTRODUCTION

As chemically interesting and bio-logically significant secondary metabolites, marine microorganisms are expected to serve as lead compounds for potential drug development or pharmacological tools for basic research in life sciences. Exploration of the marine environment began as an exciting resource during the 1960s to isolate new compounds. Much attention has been paid to identify, isolate, and purify them for useful biochemicals and utilize them in human welfare and development. Initially, the study of these natural products was focused on the chemical characterization of metabolites from macroscopic marine algae and invertebrate animals. So far, more than 3700 new natural products have been separated from these groups.

The metabolic and physiological capabilities of marine microorganisms that allow them to survive on their unique habitats also provide a great potential for production of metabolites, which are not found in terrestrial environments. However, most bioactive products of microbial origin have come from terrestrial bacteria belonging to one taxonomic group, the Actinomycetales. Although this group continues to be a prolific source of bioactive compounds, the yield of novel metabolites is decreasing and, thus, new sources of bioactive natural products need to be explored. Several natural products obtained from them have proven to be rich sources of novel compounds exhibiting many different biological activities.

THE OCEAN AS A SOURCE

Over the past decade, marine microbes have been recognized as an untapped resource for novel bioactive compounds. The ocean covers more than 70% of the earth's surface. The marine environment is vast and there is a tremendous potential to harness the ocean's organisms for their unusual metabolites. Some marine organisms are already being cultured commercially. However, mariculture practices to culture them are more practical and offers valuable source for food, pharmaceuticals, agrichemicals and other unique products. By taking this into account by volume, it represents better than 95% of the biosphere.

The marine environment ranges from nutrient-rich regions to nutritionally sparse locations where only a few organisms can survive. Many habitats in the sea have created niches for the evolution of diverse life forms . The surface and internal spaces of the sessile marine organisms provide a unique micro-habitat and several symbiotic microorganisms are reported to occupy sometimes up to 40% of their tissue volume. They are proven to be highly specific in their relationship with filter-feeding organisms like sponges, alcoynarians, ascidians, and marine plants.

The ocean environment is massively complex, consisting of extreme variations in pressure, salinity, temperature, and biological habitats. The biota of marine microorganisms has developed unique metabolic and physiological functions that not only ensure survival in extreme habitats but also offer a potential for the production of novel enzymes and bioactive metabolites for potential exploitation. It is estimated that less than 5% of the bacteria observed by microscopy are culturable under standard laboratory conditions. Out of the large number of species examined, only a fraction of marine bacteria have been isolated and cultured.

Scientists are already aware of a huge variety of marine life that is proving valuable to medical research, and still there are thousands of species not known, which hold a key promise. About one per cent of the marine animals have been examined for the pharmacological activity and only a few have been fully explored for their pharmacological properties. Various observations indicated that for every dozen organisms examined, fewer than six have been shown to have chemical and biological characteristics.

LIFE-SAVING DRUGS

The need for new therapeutic compounds has been expanding greatly because of the evolving resistance of microorganisms to existing antibiotics, the emergence of new viral diseases, and the appearance of drug-resistant tumors. Most antibiotics cannot be used in animals and humans because they are too toxic. Some toxic agents affect primarily; rapidly dividing and growing cells and have been used by adjusting dosage and directing the action of these agents in treating various diseases. Several life-saving drugs have been developed from the extracted biochemicals of terrestrial plants. The success of isolating such new drugs is immense since over 10,000 terrestrial plants have shown the occurrence of some bioactive compounds that are pharmacologically active. The work carried out on marine biota is relatively meager. Many bacteria have been recognized as a new source of therapeutic leads and have shown to produce large number of diverse biochemicals. Thus, the future of this technology is promising as a way of accessing new natural products from the marine environment.

Preliminary results suggest that marine organisms may turn out to be a more important source of anticancer compounds than terrestrial plants. Didemin, isolated from the Caribbean tunicate, Trideodemnum solidum, is active against leukemia and melanoma. It has also shown strong anti-viral and immunosuppressive activity. Manolide, a secondary metabolite isolated from the South Pacific sponge, Luffariella variabilis, and a terpene with an unusual structure has anti-inflammatory and analgesic properties besides anti-leukemic and anti-fungal properties. The sponges have proven to be a good source of secondary metabolites and several compounds have been isolated from them.

Most marine microorganisms are of the gram-negative eubacteria, cyanobacteria, and the myxobacteria groups, which are generally thought to produce many medically useful substances. Many microorganisms are capable of producing novel antibiotics. Istamycin, a new antibiotic, is formed by *Streptomyces tenjimariensis* isolated from the mud of a shallow bay in Japan. It appears to be an aminoglycoside that can inhibit a fairly large number of gram-positive and gram-negative bacteria. Aplasmomycins produced by *Streptomyces griseus* inhibit gram-positive bacteria and shows an effective antimalarial activity. *Flavobacterium ulginosum*, isolated from the seaweed of Sugami bay in Japan, produces a substance named mariactin that inhibits the development of tumors in various animals by up to 95%. Extracts of marine *Alteromonas rubre* was observed to contain rubrenoic acid, which shows bronchodilator activity. This active metabolite purified from A. rubre was not one of the many known antibiotics produced by soil microorganisms. Other biological activities observed with the culture filtrates of *Alteromonas* sp. include neuromuscular blocker, antiserotonin, anti-amphetamine, and hypotensives. Some aquatic fungal cell extracts show nicotinic blocker and polysynaptic activities.

Very few marine microorganisms are known to produce antiviral compounds. Products of the marine bacteria Vibrio marinus, inactivated enteroviruses, and *Flavobacterium* showed antiviral activity. Anti-poliovirus activity was detected with extracts of Pseudemonas and Vibrio. The antiviral effect of algae is most important as very few antiviral drugs presently exist.

Recently, studies have also suggested that some bioactive compounds isolated from marine invertebrates such as sponges, coelentrates, gorgonians, soft and hard corals, algae, mollusks, or protochordates originate from symbiotic microorganisms such as bacteria, fungi, blue green algae, dinoflagellates, or haptophytes . These compounds have shown to exhibit antibiotic, antitumor, and other pharmacological activities. Derivatives of dictyotacean algae (*Spatoglossum schmittii*) spatol demonstrated toxicity against human melanoma and astrocytomat cells in culture. Stypopodium, a potent cytotoxin produced by *Stypopodium zonale*, appears to inhibit cancer cell damage. Thus, marine microorganisms represent important and untapped resources for novel bioactive compounds.

Useful Natural Products

Several well-known chemists and biologists reviewed different aspects of marine natural products. The majority of natural products derived from microorganisms, plants, and animals could exhibit a wide range of activities. They are useful to produce important antibiotic, anti-inflammatory, antitumor, antiviral, insecticidal, and herbicidal effects. Marine toxins, enzymes, and other chemicals are also important. A relatively large number of marine organisms are known to produce secondary metabolites that possess antibiotic properties, including blue-green, green, brown, and red algae, dinoflagellates, sponges, jellyfish, sea anemones, and others apart from microorganisms. However, little research has been done on this group of organisms.

The probability of discovering new useful compounds from terrestrial microorganisms is comparatively low. Hence it is worthwhile to explore the possibilities of deriving not only new and novel antibiotics and pharmaceutically important products but also other economically valuable products such as novel enzymes, fine chemicals, and vitamins.

Green lipped mussels found in waters off New Zealand were found useful in arthritis to gain better mobility in joints. The extracts of mussels have an anti-inflammatory effect. This extract has been used safely like other conventional drugs without any side effects. At The Victoria Infirmary, Glasgow, 76% of the rheumatoid and 45% of the osteoarthritis patients reported improvement with the use of mussel extract. In India, marine mollusks and clams were examined for the recovery of heparin. Sulphated polysaccharide as a chemical present in heparin is reported to arrest bleeding from ruptured blood vessels and is also useful in the treatment of atherosclerosis. Presently, heparin is prepared from animal tissues that are rich in mast cells, such as porcine intestinal mucosa and beef lung. Naturally occurring low molecular weight heparin may be harvested more cheaply source from India's abundant coastal waters.

Marine fish oil is a rich source of eicosapentoic acid (EPA) and docosahexaenoic acids (DHA). These are particularly used in the amelioration of various cardiovascular disorders. EPA and lectins from marine species have also been used to reduce cholesterol, triglyceride, and low-density lipoprotein levels in blood and to keep the blood circulatory system normal. With anti-aggregatory properties, EPA helps in lowering blood viscosity and is used as a potent antithrombic agent. Some species examined from Indian coastal regions are rich sources of EPA and DHA and are being used in better recovery of cardiac problems. Potent medicines, prepared from such species as the air-breathing fish mudskipper (*Bolephthalmus boddeaerti*), a bivalve (*Macoma birmanica*), and a gastropod mollusk (*Telescopium telescopium*), have been marketed globally.

The horseshoe crab, (*Limulus polyphemus*) is also useful to humans. The hemolymph of the horseshoe crab is important in homeopathic therapies. Horseshoe crab extract has number of applications to treat mental exhaustion, gastro-enteric symptoms, cholera, and drowsiness after sea bathing. The amoebocyte lysate is reported to be useful in releasing all the clottable proteins or coagulation of blood by de-granulation. Traditionally the tail tip is used for healing arthritis or other joint pains in West Bengal, India. In Singapore, pregnant women eat the egg mass of the horseshoe crab to get immunity to their fetus.

Corals are known for their beauty. Some coral species have been successfully tested as bone substitutes to repair fractures in healing or to replace joints deformed by arthritis. Some species of coral resemble human bone and closely match human bone when implanted, allowing blood vessels to develop within them and acting as scaffolding for new bone to form. Some corals have just the right hole size and composition and their immune system matches with the human immune system. A team of scientists are working on the development of a synthetic material that is identical to coral for the medical replacement of bone.

Seaweed and marine algae have been valued for centuries in Asia and the Pacific Islands for their nutritional and healing properties as they are packed with potassium, vitamins A, B, C, D, and E. They are very high in iodine content and are used to treat some thyroid conditions. A brown seaweed (kelp), is found to have antibacterial and antiviral properties and has been extensively used in clinical trials to lower blood pressure in heart patients. Several marine seaweeds and submerged vegetation in seawater are reported to have antagonistic activity and have been found effective against various viruses. Purple laver contains a sulphated polysaccharide called Porphyran, which is a complex galactan, and has shown higher gelling capacity reported to inhibit the growth of Sarcoma 180 tumours in mice. A substance named porphyosin isolated from Porphyra exhibited anti-shay ulcer activity. Dinoflagellates and cyanobacteria are major sources of neurotoxins and neuropeptides that may play a vital role in understanding the development and function of the nervous system. The fluorescent pigments (phyco-bili-proteins) that occur in selected micro- and macro-algae can replace radioactive isotopes used as reporters in many immunoassays and other diagnostics. Agar polysaccharides are effective against poliovirus, herpes simplex, dengue viruses, etc. High molecular weight antitumor agents have also been isolated from marine seaweeds.

There are about 1500 species of glowing fish that create their own light. These are known as bioluminescent fish. Scientists have isolated and extracted the light-emitting genes from fish and learned to clone and produce them in labs to help medical research and diagnosis. The luminescence compound injected into a human cell can detect biological reactions happening in the human body without damaging any tissue. Clinicians are increasingly using fish luminescence in detection as radioactive compounds without any harmful effects.

ENZYMES

Marine organisms are richly endowed with many enzymes and have unique functions. This has boosted marine microbial enzyme technology in the past years, offering valuable products. These are used as pharmaceuticals, food additives, and fine chemicals. they have a diverse range of enzymatic activity and are capable of analyzing various biochemical reactions with novel enzymes. Among them, aerobic bacteria appear to be capable of producing several enzymes, such as amylases, which are well recognized to have applications in various industries. A mutant of amoeba *Trichospaherium* sp. is found to degrade and destroy plastic in the field as well as in the laboratory. Viral DNA polymerase produced by *Thermococcus litoralis*, a type of *archebacterium*, is active for more than two hours at 100°C. Enzymes isolated from extremophiles have greater stability and are longer lasting at high temperatures than those now used by industry. Their availability would open up new possibilities for efficient manufacturing.

Studies from the Orissa coast of India report that a few potent strains of marine organisms could produce novel enzymes like amalyse, protease, cellulase, *L.asperginase, acetylcholinesterase,* and *urethanase.* The results indicate that these microorganisms show potential for the production of salt-tolerant enzymes that could be used in the biotech-based food, beverage, and pharmaceutical industries. It is known that the marine environment could provide not only a wealth of newer enzymes but also many more novel enzymes by hyperproducing microorganisms, since they remain as yet an untapped resource.

Beta carotene, currently produced mainly by chemical synthesis, is used extensively as a food colorant by the margarine, cheese, soft drink, and baking industries. Studies undertaken currently indicate that beta-carotene may also be a cancer-preventing agent. Some marine bacteria belong to species of Pseudomonas. Vibrio, Aeromonas, Bacillus, and Micrococcus are capable of synthesizing vitamin B_{12}. They are reported to be about 50 to 95% of heterotrophic bacteria and are capable of secreting active thiamine. This source can be exploited for therapeutic use.

Fish Toxins

Many toxins could become valuable medicines as analgesics and muscle relaxants. Some of these have shown to have anti-tumour activity. Thus halitoxin produced by the marine sponge

Haliclona inhibits the growth of certain types of tumours. Toxins that cannot be used as medicines because of their side effects or other problems nevertheless can find use in drug design and synthesis. These are exceedingly valuable to medical research in studying nerve impulse transmission, the central nervous system, and in smooth muscle action. Tetrodotoxin, which is produced in specialized glands of the pufferfish and by certain marine bacterial species, acts to paralyze the peripheral nerves. It is being used in research to elucidate the nerve excitation mechanism. Lophotoxins, produced by gorgonia Laphogorgia, inhibits nerve-stimulated contraction of muscle.

The Future

Biotechnological prospecting of the marine environment for microorganisms is still in its infant stage. Several micro-organisms are of considerable current interest as a new promising source of metabolites and enzymes with unsuspected application potential. There is ample scope for the selection of microorganisms to produce novel enzymes. There is the possibility of producing future drugs from several marine organisms such as algae, sponges, jellyfish, corals, shark cartilage and shellfish, oyster, yeast, and fungi. A large number of toxins and poison emitted by the cone snail may be useful to prevent brain damage in animals and can also offer great promise in human health. India is blessed with a tropical climate, vast coastline, and geologically sound features that can be exploited to isolate useful microorganisms. The future for bioactive marine products for human health is bright.

A startling discovery by scientists at the Carnegie Institution puts a new twist on photosynthesis, arguably the most important biological process on Earth. Photosynthesis by plants, algae, and some bacteria supports nearly all living things by producing food from sunlight, and in the process these organisms release oxygen and absorb carbon dioxide.

But two studies 'by Arthur Grossman and colleagues+ reported in Biochimica et *Biophysica Acta and Limnology* and *Oceanography* suggest that certain marine microorganisms have evolved a way to break the rules—they get a significant proportion of their energy without a net release of oxygen or uptake of carbon dioxide. This discovery impacts not only scientists' basic understanding of photosynthesis, but importantly, it may also impact how microorganisms in the oceans affect rising levels of atmospheric carbon dioxide.

Grossman's team investigated photosynthesis in a marine *Synechococcus*, a form of photosynthetic bacteria called *cyanobacteria* (formerly blue-green algae). These single-celled organisms dominate phytoplankton populations over much of the world's oceans and are important contributors to global primary productivity. Grossman and his colleagues wanted to understand how *Synechococcus* could thrive in the iron-poor waters that cover large areas of the ocean, since certain activities of normal photosynthesis require high levels of iron. While others had suggested a potential role of oxygen as accepting electrons from the photosynthetic apparatus in place of carbon dioxide, the studies by Grossman's group show that this activity is significant in the oligotrophic (nutrient-poor) oceans, which cover about half the ocean's area.

It seems that Synechococcus in the oligotrophic oceans has solved the iron problem, at least in part, by short-circuiting the standard photosynthetic process," says Grossman. "Much of the time this organism bypasses stages in photosynthesis that require the most iron. As it turns out, these are also the stages in which carbon dioxide is taken from the atmosphere."

"We realized very quickly that there was something different about the Synechococcus that we were studying" says Shaun Bailey, the lead postdoctoral fellow working on this project. "The uptake of carbon dioxide and the photosynthetic activities didn't match, so we knew that something other than carbon dioxide was being consumed by photosynthesis, and it turned out to be oxygen." The researchers have tentatively identified the enzyme involved in this process to be plastoquinol terminal oxidase, or PTOX. They point out that this new process must be considered in understanding the net primary productivity attributed to open ocean ecosystems.

During normal photosynthesis, light energy splits water molecules. This releases oxygen and provides electrons which are then used to "fix" carbon dioxide from the atmosphere and manufacture energy-rich molecules, such as sugars. In the newly discovered process, a large proportion of these electrons are not used to fix carbon dioxide, but instead go to putting the water molecules back together, which results in much less net oxygen production.

"It might seem like the cells are just doing a futile light-driven water-to-water cycle," says Bailey. "But this is not really true since this novel cycle is also a way of using sunlight to produce energy, while protecting the photosynthetic apparatus from damage that can be caused by the absorption of light."

Capturing energy by a light-driven water-to-water cycle is critical since marine cyanobacteria are constantly using energy to acquire the meager supply of nutrients in their environment. Recently, this newly discovered phenomenon was shown to occur in nature by graduate student Kate Mackey, who made direct measurements of photosynthesis in field samples from the Atlantic and Pacific Oceans.

"The low nutrient, low iron environments account for about half of the area of the world's oceans, so they represent a large portion of the Earth's surface available for photosynthesis," says Mackey. "Our findings show that this novel cycle occurs in two major ocean basins and suggest that a substantial amount of energy from sunlight gets re-routed away from carbon fixation during photosynthesis. This may mean that less carbon dioxide is being removed from the atmosphere by the open ocean photosynthetic organisms than was previously believed."

"This discovery represents a paradigm shift in our view of photosynthesis by organisms in the vast, nutrient-starved areas of the open ocean", says Joe Berry of the Carnegie Institution's Department of Global Ecology. "We had assumed that like higher plants, the goal was to make carbohydrates from carbon dioxide and store them for later use as a source of energy for any number of cellular functions or growth. We now know that some organisms short-circuit this complicated process, using light in a minimalist way to power cellular processes directly with a far simpler and cheaper (in terms of scarce nutrients such as iron) photosynthetic apparatus. We don't know the full significance of this finding yet, but it is certain to change the way we interpret optical measurements of photosynthetic pigments in the ocean and the way we model ocean productivity."

Wolf Frommer, director of the Carnegie Institution's Department of Plant Biology, agrees about the discovery's ground-breaking importance:

"If we thought we have understood photosynthesis, this study proves that there is much to be learned about these basic physiological processes. The findings of Grossman's laboratory together with previous evidence reported by Greg Vanlerberghe from the University of Toronto showing that the gene encoding PTOX appears to be widespread in marine cyanobacteria will add depth and a mechanistic foundation for the modeling of primary productivity in the ocean."

New Centres for Oceans and Human Health

One in twenty: that's the percentage of people who develop rashes, nausea, or other symptoms after swimming in marine waters with levels of pollution deemed acceptable by the European Union and the U.S Environmental Protection Agency, according to the World Health Organisation. Most people infected by pathogens in the water generally will recover quickly, as will those exposed to harmful algal blooms (HABs), sometimes known as red tides. But in rare cases, especially among the young or the elderly, illnesses can be severe, resulting in life-threatening infections or—in the case of exposure to algal toxins—paralysis, neurological disorders, and death.

With the aim of reducing disease and death among people who eat seafood and swim in oceans and bays, scientists at four new research centers will study the ecology, oceanography, genetics, and health effects of marine pathogens. Center scientists will also search for new marine sources of antibiotics and other drugs. The National Science Foundation (NSF) and the NIEHS announced funding for the four joint Centres for Oceans and

Human Health on 22nd April, 2004. The two agencies will invest about $5 million annually for a total of $25 million to support the centres, which are located at the University of Washington, the University of Hawaii, Woods Hole Oceanographic Institution, and the University of Miami.

"Oceans have become conduits for a number of environmental threats to human health," said NIEHS director Kenneth Olden at the time of the centers' announcement. "At the same time, oceans harbor a diverse array of organisms that show great promise for providing new drugs to combat cancer and fight infectious diseases. In order to guard against health threats and to take advantage of medicinal benefits that oceans might provide, the impact of oceans on human health must be more fully explored."

The NIEHS has recognized the linkage between oceans and human health for a number of years, as evidenced by its support of marine-related research projects on the effects of toxins from HABs and contaminated seafood consumption, and a set of Freshwater and Marine Biological Research Centers," says center programme administrator Fred Tyson. "Moreover, the NIEHS has embraced the concept of using multi- and interdisciplinary approaches to address complex environmental health issues with programmes such as its Centers for Children's Environmental Health and Disease Prevention Research and Centres for Population Health and Health Disparities.

HARMFUL ALGAL BLOOMS

No one is sure why algae rapidly increase in abundance, or "bloom," and why some blooms produce toxins and others do not. Gaining a better understanding of the factors that cause HABs is a primary research focus of all four centers. Each center will study conditions, such as currents, salinity, water temperature, and nutrient loading, that may trigger blooms and the release of toxins. Researchers will also study blooms at the microscopic level, using genetic analysis to identify the different species and subspecies that bloom under various conditions. The ultimate goal of such research is to find better ways to predict blooms and detect the toxins they produce, thus reducing human exposure and disease. For example, the results of genetic studies could be used to develop molecular probes to detect toxic species and prevent exposure.

Once toxins accumulate in the water during certain blooms, swimmers and beach goers can be affected through skin contact or by swallowing water or inhaling aerosolized spray. However, eating seafood is the most common route of exposure. Algal toxins work their way up the food chain by accumulating in the digestive tracts, and sometimes the muscle tissue, of fish and shellfish. In the United States, few people eat tainted shellfish, thanks to monitoring programmes that close commercial and recreational shellfish beds if toxins are detected. However, there are serious gaps in toxin protection programmes for tropical and subtropical areas.

One reason is a lack of programmes that regularly test for the ciguatera toxins found in tropical reef fish. Another is the high cost of most existing testing programmes for all types of HABs. "We need to develop low-cost, effective monitoring protocols affordable by poor populations in tropical and subtropical regions," says Lora Fleming, director of the University of Miami's Center for Subtropical and Tropical Oceans and Human Health. "HAB poisoning is a problem in these areas because poverty precludes expensive monitoring." For example, in a case in Guatemala, 187 people contracted paralytic shellfish poisoning, and 26 died, including half of the children who were exposed. Further, says Fleming, organisms and their associated toxins and diseases can turn up in new geographic locations and new transvectors. While this can happen in developed nations, too, those countries can better afford to evaluate the outbreaks that result and sometimes even prevent them. Many existing tests for algal toxins use mouse bioassays, which are costly and time-consuming.

Another issue with current testing is that existing monitoring programmes are designed to detect relatively high toxin levels that cause noticeable illness. Don Anderson, a senior scientist at Woods Hole, says harvesting closures typically occur only after toxins reach certain threshold concentrations, such as

80 micrograms of saxitoxin per 100 grams of meat, or 20 micrograms of domoic acid per gram of meat. "It is therefore possible for products to be sold and consumed that contain low levels of toxins that are detectable, but below these thresholds," he says. "We do not know the effects of long-term exposure to these low levels, especially in sensitive segments of the population, such as the elderly and children."

The interdisciplinary center teams will study several types of algal toxins that cause human illness. Researchers at the Woods Hole Center for Oceans and Human Health will focus on saxitoxins, a group of more than 20 neurotoxins that cause paralytic shellfish poisoning. Saxitoxins attack the nervous system, causing numbness, dizziness, headache, and in the worst cases, respiratory failure. In the oceans, they are produced by dinoflagellates, single-celled organisms distinguished by dual flagella, or hair-like structures, that propel them through the water.

The Woods Hole team will focus on two species of algae: *Alexandrium fundyense,* which produces *saxitoxins,* and *A. ostenfeldii,* which also generates spirolides, a type of neurotoxin first detected off Nova Scotia in 1995. Spirolides kill mice quickly. However "there is as yet no [human] poisoning syndrome linked to spirolides, and thus far, no monitoring for spirolides anywhere in the United States.

Efforts to predict or forecast toxicity in the future will have to recognize that we are not just dealing with a single species but rather a complex of strains or subpopulations that will respond differently to the environment. For example, in the Woods Hole researchers' study area off the coast of Maine, Anderson predicts that researchers will find multiple varieties of Alexandrium that vary significantly in toxicity, physiology, and behaviour—for example, some will grow faster, others will swim faster, still others may migrate vertically in different patterns.

Researchers at the University of Miami will study a subtropical species of dinoflagellate (Karenia brevis, formerly Gymnodinium breve) that produces brevetoxins, the cause of neurotoxic shellfish poisoning. Neurotoxic shellfish poisoning is similar to paralytic shellfish poisoning, but with somewhat milder symptoms. With [neurotoxic shellfish poisoning], the primary concern used to be eating seafood. But now we have concerns about skin contact and drinking water. There is also evidence that the toxins can be aerosolized, because people have developed asthma-like symptoms while simply walking on the beach and, presumably, inhaling minute amounts of sea spray.

Researchers at the University of Hawaii are studying another warm-water dinoflagellate (Gambierdiscus toxicus) that produces the ciguatoxins that cause ciguatera fish poisoning. People are most often exposed by eating reef fish such as barracuda and grouper. Symptoms—which can include gastrointestinal distress, blurred vision, irregular heartbeat, depression, and the reversal of temperature sensations (hot things feel cold and vice versa)—can last months after exposure. Like other diseases caused by algal toxins, ciguatera fish poisoning is underreported. According to the Centers for Disease Control and Prevention, perhaps 50,000 people worldwide suffer from ciguatera, but only 2-10% of U.S. cases are reported.

The Hawaii team will work on molecular probes and markers with the goal of eventually developing monitoring systems that could be used throughout Pacific and Caribbean waters. Reef fish usually are not migratory, so specific reefs could be monitored and marked the same way shellfish beds are to reduce exposures. Fishermen tell us they don't fish certain reefs because there's ciguatera there. An added benefit to ciguatera reef closures, she adds, is that they could protect reef fish from overharvest.

Domoic acid caught the attention of medical and oceanographic researchers when it was found in mussels that sickened more than 100 people and killed 3 in a 1987 outbreak in Canada's Prince Edward Island. Once a test for domoic acid was developed, it was detected in razor clams in Washington State and Dungeness crabs in Alaska. Domoic acid causes amnesic shellfish poisoning, a neurological illness with symptoms ranging from gastrointestinal distress to permanent loss of short-term memory. Birds and mammals can also be affected. Researchers now suspect that a 1961 incident in Capitola, California, in which nonviolent seabirds attacked people and dive-bombed windows (later inspiring the Alfred Hitchcock thriller The Birds) was caused when the birds ate fish tainted with domoic acid. Many recent seabird and marine mammal mortalities on the East and West coasts have been linked to domoic acid moving through the food web, even to whales.

Researchers at the University of Washington Pacific Northwest Center for Human Health and Ocean Sciences, directed by Elaine Faustman, will study environmental factors associated with blooms of single-celled plankton called diatoms (Pseudo-nitzschia spp.) that produce domoic acid. Studies will compare conditions in the open Pacific with the more enclosed, estuarine waters of Puget Sound, where domoic acid was detected in 2003. The researchers will also analyze the molecular mechanisms by which domoic acid kills nerve cells, as well as examine possible reasons why some shellfish, such as razor clams, retain domoic acid for weeks or months, while other species eliminate the toxin quickly.

Of particular concern to these researchers is determining who is at greatest risk of neurological damage from exposure to domoic acid. Susceptible individuals may include members of Native American tribes and Asian/Pacific Islander groups, who tend to eat much more seafood than other groups and who sometimes eat parts of shellfish where the toxin accumulates, such as the hepatopancreas. In addition, children and the elderly are potentially more vulnerable to the effects of domoic acid and other algal toxins.

Parasitic Protists and Bacteria

In addition to algal toxins, which cannot be passed from one person to the next, center researchers will also study infectious agents of disease. Researchers at Woods Hole will study the prevalence and ecology of a range of bacteria and parasitic protists including amoebae in temperate waters, while the centers at Miami and Hawaii will focus on developing better indicators for infectious pathogens in warmer waters.

The Woods Hole researchers will study Mount Hope Bay in Massachusetts, where warm water discharged from a power plant may encourage the growth of pathogens that would not normally be found there. Many pathogens enter marine waters, especially through sewage, but "whether they survive marine conditions or accumulate in concentrations that could be vectored back to humans is a question."

The group will study the ecology of several types of bacteria in the same genus (Vibrio) as the species that causes cholera. They will also hunt for rarer organisms, including amoebae that harbor the bacterium that causes Legionnaire disease (*Legionella pneumophila*) and a parasitic amoeba (*Naegleria fowleri*) that thrives in warm water and travels up the noses of swimmers and divers to cause rare, but often fatal, infections of the nervous system and brain.

Researchers at the Hawaii and Miami centres will search for microorganisms that can serve as better indicators of marine pathogens. A problem is that a lot of the indicators commonly in use have nonfecal sources, or they aren't appropriate for tropical or subtropical waters. He adds that recent studies have shown that Enterococcus species—the standard indicator recommended by the U.S. Environmental Prevention Agency—grow naturally in the soil of Hawaii, and probably other tropical areas, and so don't necessarily signal fecal pollution.

PHARMACEUTICALS FROM THE SEA

At the same time as the center researchers study marine organisms that harm, they will also prospect for those that heal. "Scientists believe the oceans represent a promising source of novel compounds with therapeutic and/or disease-fighting capabilities," Hollings said in his 24 March 2004 address to the U.S. Senate. "At present, there are only three marine compounds in clinical use—and these were developed in the 1950s. While there are some new compounds in the pipeline, we need to speed research efforts to ensure we get more products approved."

In their search for pharmaceuticals, center researchers at the University of Hawaii will focus on marine microbes. The fact is that there are more microbes in the ocean than there are stars in the sky. It's reasonable to assume that some have devised biochemical defenses to protect themselves from potential enemies, and that some of those biochemical defenses may involve compounds of interest to humans.

The Hawaii team will screen microbe extracts for tumor-fighting and antibiotic properties, working in collaboration with the University of Hawaii Cancer Research Centre and private laboratories. Given recent results from other laboratories, it's possible that center researchers may find compounds in both categories. Researchers affiliated with the National Cancer Institute and other groups have extracted compounds from marine sponges and corals that inhibit the growth of tumor cells, and a group from the Scripps Institution of Oceanography has isolated marine bacteria closely related to the terrestrial organisms from which antimycin antibiotics were originally derived.

MICROBES IN THE OCEAN

The oceans teem with microorganisms such as bacteria, viruses, and protists. Many of these microbes fundamentally influence the ocean's ability to sustain life on Earth. Some microbes living and transported in ocean water, however, threaten human health.

Occurrence and Role in the Ocean

In the open ocean, far from the influences of coastal human habitation, sea water still contains huge numbers of microbes. Coastal areas can contain even greater concentrations. Vast numbers of bacteria and plankton occur both at the surface and in deep ocean waters. Viruses are entities that require bacteria or other cells in order to make copies of their genetic material and to construct new casings that house the genetic material. Scientific studies have shown that 10 to 100 million viruses can be present in a teaspoonful of sea water.

Plankton

More plankton exist in sea water than any other organism. Microscopic forms include protists and bacteria. Phytoplankton are photosynthetic organisms, including algae. By harvesting the energy of the Sun and converting it to their tissues, phytoplankton form the basis of the food chain in the ocean. All ocean organisms depend on phytoplankton either directly or indirectly. Eventually, humans consume ocean creatures such as fish. Even human life, therefore, is tied to the presence of phytoplankton.

Microbes such as plankton also have other benefits. In the ocean, they help make some nutrients available to other living marine creatures. Elemental iron, for example, is important for living creatures but is scarce in the ocean. Sunlight can change iron to a form that can be taken up by plankton and other microbes. The microorganisms are used as food by other organisms, such as fish and ocean mammals, making the iron available to other creatures in the food chain.

Viruses

Marine viruses can be both detrimental and beneficial to the ocean's health. Some viruses attack and kill plankton, eliminating the base of the ocean food chain in a particular area. At the same time, this dead plankton can become a source of carbon that is not otherwise readily available to other sea life. Scientists estimate that up to 25 per cent of all living carbon in the oceans is made available through the action of viruses. When these aspects remain in proper balance, the ocean functions normally.

Bacteria

Bacteria are single-celled organisms without cell nuclei. They are found in all portions of the water column, the sediment surface, and the sediments themselves. Some are aerobic (requiring oxygen), whereas others are anaerobic (not requiring oxygen). Most bacteria are free-living, but some live as partners (symbionts) within other organisms. For instance, many deep-sea fish harbor symbiotic bacteria that emit light, which the fish use to signal other members of their species. The bacteria's ability to emit light is called bioluminescence. Bioluminescence causes water to glow, a phenomenon most noticeable at the surface but present at all depths.

Cyanobacteria, a type of bacteria, played an important role in the history of Earth and in ocean processes, including the development of stromatolites. Living in colonies, the cyanobacteria produced oxygen during the process of photosynthesis, which generated the oxygen in the Earth's atmosphere that many living beings require today. Although cyanobacteria also are called "blue-green algae", it is important to remember that cyanobacteria are relatives of bacteria and not algae. They are, however, related to the chloroplasts within algae; the chloroplasts used by some plants to produce food are actually cyanobacteria living within plants' cell walls.

Some marine bacteria can interact with diatoms, another type of marine microbe, in such a way that influences the cycling of silicon in the ocean. Diatoms, a group of unicellular algae, are characterized by their highly ornate two-part shell-like structures made from silica.

One species of bacteria, Thiomargarita namibiensis, plays a critical role in hydrogen sulfide eruptions from diatomaceous sediments off Africa's Namibia coast. Known as the "sulfur pearl of Namibia," this anaerobic species digests organic matter under low-oxygen (or no-oxygen) conditions that are caused by high rates of phytoplankton growth in the Benguela upwelling zone, and the subsequent decay of large masses of dead phytoplankton that have fallen to the seafloor.

The anaerobic activity leads to the formation of hydrogen sulfide gas (H_2S) in the sediment. Over time, the gases build up and are periodically released into the water column in a "sulfide eruption." At the water surface, the H_2S oxidizes to microgranules of sulfur, discoloring it a milky green. These surface features can be observed by satellites.

Recent Discoveries: Archaea

Knowledge of the diversity of microbial life in the oceans continues to grow. Until the 1990s, knowledge of microbial populations was determined using assays that relied on the growth of the microbe. Now, detection and identification of microbes are possible by the examination of their genetic material. These molecular assay techniques have revealed much larger numbers and types of microbes in the ocean than scientists previously suspected.

The bacteria-like microbes known as Archaea represent one example of research surprising to marine microbiologists. Archaea are one of the major domains of life on Earth. Since their discovery in 1970, these microorganisms have been found in many extreme environments on Earth, including hydrothermal vents on the ocean floor. Recently, scientists determined that Archaea also exist in the open sea. Moreover, these microbes may comprise up to half the mass of life in the oceans, and so must play an important role in the processes that occur in the oceans.

Since 2001, the examination of sediment from the sea bottom has revealed the presence of another type of Archaea that exist by using methane, an important gas that is contributing to the warming of the Earth's atmosphere. In addition, bacteria that exist by using the rocks of the sea bottom as food have been discovered. The release of material from the sediment by the action of the rock-using bacteria may influence the chemistry of the oceans.

Animal and Human Health Impacts

Some ocean microorganisms can cause unhealthy effects in both land and sea animals. This is particularly true in coastal regions, where the influence of humans is more evident. For example, runoff or the deliberate release of sewage into the oceans releases huge numbers of bacteria and viruses into the water. These microbes normally live in the intestinal tracts of humans and other warm-blooded animals. The water that is contaminated by these microbes can be the source of diseases.

Just as humans are susceptible to microbial infections, so too can marine animals (e.g., mammals) develop infections. It is believed that infections are to blame for at least some cases of marine mammal "beachings," in which whales or dolphins become stranded on the shore. It is thought that human pollution may exacerbate this problem by increasing the likelihood of infection and decreasing the quality of the water.

Another potentially harmful microbe found naturally in the ocean is a protist called a dinoflagellate. At certain times and under certain conditions, some dinoflagellate species and other algal species can undergo population explosions called blooms, sometimes in response to human-caused pollution. These blooms often are called "red tides" because the algal pigments color the water. Further, some of the bloom-causing algae may produce natural poisons known as biotoxins. These biotoxins are transferred to ocean animals that feed on the toxin-containing algae, and also are released into the water as the dead algae decay. These biotoxins can bioaccumulate in the ocean food chain, sickening or killing higher-order animal consumers and tainting fisheries and shellfisheries used by humans.

Human Pathogens

The ocean has long been used as a means of disposal of human wastes, and still is used this way in many areas of the world. The enormous size of the oceans historically and incorrectly was assumed able to dilute any noxious material to concentrations low enough to render them harmless. While small amounts of sewage can dissipate quickly, the routine dumping of large amounts of human waste has caused long-term harm to many shoreline ecosystems, and rendered many coastal waters unfit for recreational activities.

For example, high numbers of disease-causing viruses and a bacterial species called *Escherichia coli* can occur in coastal waters influenced by human wastes (e.g., sewage). If ingested, these microbes can cause intestinal upset and organ damage that varies from inconvenient to life-threatening. Furthermore, the viruses that cause hepatitis A, hepatitis E, and polio, along with protozoan pathogens that cause giardiasis and cryptosporidiosis, are sometimes found in coastal waters and estuaries.

Another example related to improperly treated or untreated sewage is *Vibrio cholerae*, the bacterium that causes the intestinal infection cholera. Huge outbreaks of cholera have occurred throughout recorded history. In recent years, severe outbreaks of cholera occurred in Bangladesh and in Peru. Some evidence suggests that the cyclical temperature increase associated with El Niño currents increases the likelihood of such outbreaks, by increasing the ability of the bacterium to reproduce in polluted waters.

Vibrio cholerae and other infectious bacteria and viruses also are commonly present in high numbers in the ballast water of ocean-going ships, and can be widely spread by these ships. Ballast water is water that is pumped into the hull of a ship in order to stabilize it against the rough conditions of an ocean voyage. When the ship reaches port, millions of liters of ballast water are pumped out into the harbor, releasing the microbes into the new environment. Currents can act as highways for microbes, carrying them along in the water flow. Viruses, which can live in the ocean for days or weeks, can be transported great distances via ocean currents.

6

Harmful Algal Blooms

INTRODUCTION

Single-celled algae are almost always present in sea water even if the water looks clear. When high concentrations of certain species of dinoflagellates are present, patches of water look red because these algae contain red pigments—hence the name "red tide." High concentrations of other algae may turn sea water orange, yellow, brown, or purple. Red tides have been witnessed for centuries and have been seen all over the world.

Dense concentrations of algae are referred to as blooms, because the algae have multiplied rapidly to become concentrated in high numbers. In a bloom, there could be tens of millions of cells in a liter of sea water. Most blooms are not harmful, but some have the potential to be harmful, whether by virtue of natural biotoxins (poisons) produced by certain species of algae, or by the oxygen-depleting process initiated upon the death and subsequent decay of large concentrations of algae.

Harmful algal blooms, or HABs, cause millions of dollars in damage when there are massive fish kills to be cleaned up, beaches declared offlimits, fisheries and shellfisheries closed to harvesting, and medical treatment provided for people poisoned by marine biotoxins in the seafood they ate. Many scientists believe that harmful algal blooms are becoming more prevalent, but they point out that increased monitoring efforts are detecting more occurrences.

MECHANISMS OF HARM

How do certain microscopic algae—that is, types of phytoplankton—cause harm to fish, shellfish, marine mammals, seabirds, and people? Basically, there are four ways.

First, the physical presence of so many cells may suffocate fish by clogging or irritating the gills. Second, when the densely concentrated algal cells die off, the decay process, assisted by bacteria, can deplete the water of oxygen, which in turn can lead to the death of oxygen-dependent marine creatures. (Algae, being plants, require nutrients such as nitrogen and phosphorus to grow. When they have used up the nutrients, they tend to die off all at once.) Such oxygen-related impacts are most visible in shallow bays, inlets, or seas.

Third, some algal species produce deadly toxins which directly kill the animals that ingest the poisons. Dinoflagellate toxins have killed mussels, abalone, and fish. Airborne toxins (i.e., toxins that are aerosolized) have caused respiratory problems and eye and skin irritation to people along beaches where harmful algal blooms were present.

Fourth, shellfish such as mussels, clams, and oysters feed by filtering particles, including phytoplankton, from sea water. Toxins from certain dinoflagellate or diatom species accumulate in the tissues of shellfish. When people, sea mammals, or seabirds eat the shellfish, they ingest the toxins as well.

There are different kinds of toxins that cause different kinds of symptoms which, in humans, typically are neurological. Some toxins are deadly. The table above shows the categories of human poisoning; the organism associated with the toxicity; and the causative toxin (sometimes among a suite of toxins).

Types and Examples of Toxicity

The algal species that can be toxic in some circumstances are not always toxic. When they are toxic, they cause harm by being eaten by larger organisms. As a toxin is passed up the food chain, it becomes concentrated in larger and larger animals such as fish and shellfish, and eventually is ingested by people who eat seafood containing the toxin. In the following overview, only two examples of the various algal species known to cause toxicity are discussed.

PFIESTERIA

An exception to toxin transmission up the food chain is the dinoflagellate Pfiesteria. Instead of being eaten, it does the eating—usually small organisms but also fish. It uses its toxins to make the fish lethargic and to injure the fish's skin.

Toxins can also get into the air and cause harm to people, as happened in 1991 when this peculiar organism was first discovered in a laboratory in North Carolina. Later it was found in connection with fish kills in the Albemarle–Pamlico estuary and other estuarine environments on the U.S. Atlantic and Gulf of Mexico coasts. Blooms of the more common dinoflagellate, Gymnodinium breve, when present in nearshore waters, can be picked up by surf and wind and carried to the seashore in the air. The microscopic organisms cause skin and eye irritation in people exposed to this toxic aerosol.

A red-tide dinoflagellate typically has a vegetative stage, in which it multiplies by cell division, and a cyst stage, in which two cells combine to form gametes enclosed in a cyst (a type of covering). The cysts sink to the seafloor until conditions are favorable for a return to the vegetative stage.

The Pfiesteria organism has a minimum of twenty-four stages in its life cycle, of which at least four are toxic. The life stages include flagellated cells that swim in the water, amoeboid forms both in the water and in bottom muds, and cysts that rest on the bottom. The different forms vary in size from too small to see with an ordinary microscope, to a speck visible to the naked eye.

Most of the time, the Pfiesteria dinoflagellate is a nontoxic predator that feeds on small organisms such as algae, bacteria, and small animals. It becomes toxic when cyst forms detect fish excretions or secretions. Encysted cells emerge and become toxic. They damage the fish with the toxin, then feed on the epidermal tissue, blood, and other substances that leak from sores on the incapacitated fish. When the fish are dead, the cells change to the amoeboid stages and feed on the fish carcass.

Pseudo-nitzschia

Diatoms are single-celled marine or fresh-water algae that have shell-like structures, called frustules, made of silica. Until recently, diatoms were not associated with biotoxin poisonings. But in 1987, an outbreak of domoic acid poisoning was reported in Canada. The domoic acid came from a diatom, Pseudo-nitzschia. Domoic acid has caused permanent memory loss and death in humans.

In 1991, examination of the stomach contents of dead seabirds found along the beaches of Monterey Bay, California revealed high levels of domoic acid. The birds had been eating anchovies which had been consuming Pseudo-nitzschia. In 1998, sea lion deaths on the California coast also were associated with domoic acid, which entered the food chain via toxigenic diatoms, which were eaten by anchovies that in turn were eaten by the sea lions. To prevent human illness when such toxicity is found, state health departments temporarily close beaches and federal regulatory agencies temporarily close fisheries and shellfisheries along the affected coast.

Human Impacts and Intervention

Although the economic impacts of HAB outbreaks has not been quantified on a national basis, the direct and indirect costs to even a single fishery closure can reach millions of dollars. In addition to loss of revenue to fish and shellfish industries, there are impacts on recreational fishing and tourism and their associated businesses.

Harmful algal blooms also can threaten the aquaculture industry. For example, unpredictable and destructive blooms of the small flagellate Heterosigma have threatened the commercial farmed salmon industry in Washington state (USA) and British Columbia (Canada). Heterosigma blooms also have destroyed some captive populations of threatened and endangered salmon being raised in net pens before their release to the wild.

All of the U.S. coastal states have developed monitoring programmes with regular testing of fish and shellfish from beaches. Officials and volunteers watch the shores for patches of colored water, fish kills, the beaching of marine mammals and other unusual activity, or reports of human illness following consumption of fish or shellfish. When toxins show up in laboratory analyses of samples of edible species, warnings are issued and shellfish harvesting and some kinds of fishing may be halted. Economic losses can be high when commercial fishing and aquaculture operations (including fish and shellfish farms) are affected.

To better manage the human risk associated with HABs, scientists are continuing to research methods of rapid analysis to identify toxic phytoplankton species and to detect marine biotoxins in water, phytoplankton, and animals. Better monitoring can help decrease the incidence of overly conservative fishery closures by delineating the extent of the threat, thus reducing the need for broad-scale closures due to lack of information.

The Harmful Algal Bloom and Hypoxia Research and Control Act was enacted in 1998. The Act recognizes that HABs threaten coastal ecosystems and endanger human health. A national assessment, published in early 2001, recognized the threat to human health and coastal economies, but found that management options are limited. HAB impacts can be minimized through monitoring programmes that regularly sample shellfish to detect HAB toxins, and issue warnings when toxins are found. Satellite remote sensing can track offshore blooms, alerting coastal communities to potential problems as blooms come inshore.

Algal Blooms in the Ocean

The ocean, that vast body of water covering 71 per cent of the Earth's surface, is divided into four major basins: the Pacific, Atlantic, Indian, and Arctic Oceans. These large basins are interconnected with various shallow seas, such as the Mediterranean Sea, the Gulf of Mexico, and the South China Sea.

Oceans and seas abound with life, ranging from microscopic unicellular (one-celled) organisms to multicellular (many-celled) animals. Algae is an important life form in the ocean. Life in the ocean is maintained in balance by forces of nature and by predator–prey relationships, unless some external pressures upset the balance. When a balance upset leads to conditions more favorable for the reproduction and growth of algae, an explosive increase in the number of algal cell density occurs. Such rapid increases in the algae population are called algal blooms.

During a bloom, a liter of water may contain millions of algae. The most widely publicized type of algal bloom is associated with species that produce a toxin (chemical substance) harmful to animals that feed on the algae (and hence is known as a harmful algal bloom), and/or algae that cause a tint in the water because of the photosynthetic pigments they contain. The latter commonly is known as a "red tide," but different pigments can turn the water red, brown, purple, orange, or yellow. Depending on the circumstances and the species present, a red tide may or may not be harmful. Although not all algal blooms in the ocean produce highly visible effects nor are all blooms harmful, they nonetheless affect life in the ocean and on land in both beneficial and harmful ways.

REQUIREMENTS FOR A BLOOM

Algae require warmth, sunlight, and nutrients to grow and reproduce, so they live in the upper 60 to 90 meters (200 to 300 feet) of ocean water. The upper layer of water, the epipelagic zone, is rich in oxygen, penetrated by sunlight, and warmer than water at lower levels. As algae and other organisms that live in the ocean die, they fall to the bottom of the ocean, where they decay and release the compounds from which they were made. Under certain conditions, these nutrients can deplete the oxygen in the water.

Temperature and salt concentration determine the density of water and how water moves (currents). Cold water is denser (heavier) and sinks from the surface (downwelling). Other water moves across to replace it. Eventually, water at the surface is replaced by water that has risen, or upwelled, from the bottom to the surface somewhere else in the ocean. These upwellings bring nutrient-rich waters to the top. This increase in nutrients can trigger algae blooms.

An increase in nutrients also may be caused by activities of humans, such as runoff from animal farms or fertilized croplands and lawns, or atmospheric deposition of sulfur and nitrogen compounds or oxides derived from the burning of fossil fuel. These nutrients lead to blooms in coastal waters to a greater extent than in the open ocean.

However, some of these nutrients do find their way to the open ocean far from shore, and contribute to the formation of blooms in the open ocean. Their movement is aided by the wind and by ocean currents. Algae blooms in the open ocean are not usually harmful; instead, they provide many benefits, largely deriving from the fact that the open ocean is relatively unproductive (low in nutrients).

Algae and Photosynthesis

Algae are referred to as plants because, like plants, they produce organic compounds from inorganic compounds (carbon dioxide and water) by capturing and using the energy from sunlight. Most algae are eukaryotic, an exception being the blue-green algae (cyanobacteria).

Photosynthesis takes place in organelles called chloroplasts in eukaryotic, cells. Chloroplasts contain an outer and an inner membrane and pancakeshaped structures called thylakoids. Energy is captured from sunlight by pigments (chlorophylls a and b and carotenoids) stored in the thylakoids.

Photosynthesis occurs in two stages commonly referred to as the light reactions (light is required) and the dark reactions (no light is directly required). During the light reactions, energy captured from sunlight is used to split (dissociate) water molecules. Electrons released from this reaction are passed down a series of electron carrier molecules, leading to the storage of the energy in the form of ATP (adenosine triphosphate). This is the form in which living organisms store energy to be used immediately for carrying out chemical reactions and other activities.

Oxygen is produced as a byproduct of the light reactions. During the dark reactions (Calvin cycle), six molecules of carbon dioxide are used to make sugar (glucose). Because algae use carbon dioxide and release oxygen as a product of the light reactions, these plants play an important role in maintaining the proper concentrations of carbon dioxide and oxygen in the environment, via the carbon cycle and oxygen cycle.

Algae, like green plants, produce the first organic compounds in the food chain and thus are referred to as primary producers. Other organisms cannot use inorganic molecules to make the organic compounds that they need for life, and therefore depend on algae and other plants as the initial source of organic compounds. These organisms either eat algae to obtain organic compounds, or obtain them from the water when they are released after the algae die.

Types of Algae that May Bloom

Cyanobacteria

Cyanobacteria, also known as blue-green algae, are one of the oldest known types of algae and are believed to have played a major role in the addition of oxygen to the Earth's early atmosphere. Some cyanobacteria carry out nitrogen fixation, which is the conversion of nitrogen gas into nitrogen compounds that can be used by other primary producers.

Diatoms

Diatoms are unicellular and have a cell wall composed of silica, a glass-like material, which comprises a shell-like structure called a frustule. When diatoms die, the frustules settle to the bottom of the ocean floor and combine with the soil to form diatomaceous earth. Diatomaceous earth is used in products such as filters for swimming pools, as temperature and sound insulators, and as an abrasive in toothpaste.

Dinoflagellates

Dinoflagellates have two unequal flagella that help them direct their movement. Many of these organisms contain colored pigments that cause the water to appear colored when these organisms bloom, leading to the terms "red tide" or "brown tide," for example. Some dinoflagellates live in close association with marine animals, such as sponges, sea anemones, giant clams, and corals. The golden-brown photosynthetic cells found in these animals, called zooxanthellae, actually are dinoflagellates.

Coccolithophores

Coccolithophores are cells covered with button-like structures called coccoliths made of calcium carbonate. The coccoliths give the ocean a milky white or turquoise appearance during intense blooms. The long-term flux of coccoliths to the ocean floor is the main process responsible for the formation of chalk and limestone. Coccolithophores and some other algae participate in the sulfur cycle and produce the gas dimethyl sulfide. This is the primary way that sulfur is carried between ocean and land. Dimethyl sulfide leaves the surface of the water and reacts with oxygen in the atmosphere to form tiny sulfuric acid droplets. These droplets are carried over land and fall back to land in the form of precipitation. They also aid in the formation of clouds, which partially block the transmission of harmful ultraviolet light that penetrates the surface water. Cloud formation also is thought to encourage surface winds that promote the movement of surface water, leading to upwellings that bring nutrients to the surface.

BENEFITS OF ALGAL BLOOMS

Algal blooms provide large concentrations of algae that produce organic compounds needed by higher organisms, ranging from oysters, clams, and mussels to human beings. For this reason, productivity increases in areas where algal blooms occur. More algae in the water means that more carbon dioxide is used from the atmosphere and that more oxygen is released into the atmosphere. Oxygen is necessary for many living things, including humans. As noted previously, the production of dimethyl sulfide gas helps protect algae from harmful ultraviolet rays so they remain healthy and thus are able to continue the cycle of sustaining life on Earth. Even in the coldest parts of the ocean, algae provide the primary source of organic material to animals at the bottom of the food chain. Organic materials are moved up the food chain as higher organisms feed on those lower down the chain. For example, algae have been found in Antarctic sea ice. As sea water freezes, algae living in the water are frozen in the ice, where they later can be released during a thaw. These algae are a vital source of food for krill, the shrimp-like organisms eaten by penguins, seals, seabirds, and whales.

Algal Blooms in Fresh Water

Aquatic ecologists are concerned with blooms (very high cell densities) of algae in reservoirs, lakes, and streams because their occurrence can have ecological, aesthetic, and human health impacts. In waterbodies used for water supply, algal blooms can cause physical problems (e.g., clogging screens) or can cause taste and odor problems in waters used for drinking. Blooms involving toxin-producing species can pose serious threats to animals and humans.

Algae in Aquatic Ecosystems

The term "algae" is generally used to refer to a wide variety of different and dissimilar photosynthetic organisms, generally microscopic. Depending on the species, algae can inhabit fresh or salt water. In modern taxonomic systems, algae are usually assigned to one of six divisions. The misnamed blue-green algae are often grouped with algae because of the chloroplasts contained within the cells. However, these organisms are actually photosynthetic bacteria assigned to the group cyanobacteria. Fresh-water algae, also called phytoplankton, vary in shape and color, and are found in a large range of habitats, such as ponds, lakes, reservoirs, and streams. They are a natural and essential part of the ecosystem. In these habitats, the phytoplankton are the base of the aquatic food chain. Small fresh-water crustaceans and other small animals consume the phytoplankton and in turn are consumed by larger animals.

BLOOM OCCURRENCES AND IMPACT

Under certain conditions, several species of true algae as well as the cyanobacteria are capable of causing various nuisance effects in fresh water, such as excessive accumulations of foams, scums, and discoloration of the water. When the numbers of algae in a lake or a river increase explosively, an algal "bloom" is the result. Lakes, ponds, and slow-moving rivers are most susceptible to blooms.

Algal blooms are natural occurrences, and may occur with regularity (e.g., every summer), depending on weather and water conditions. The likelihood of a bloom depends on local conditions and characteristics of the particular body of water. Blooms generally occur where there are high levels of nutrients present, together with the occurrence of warm, sunny, calm conditions. However, human activity often can trigger or accelerate algal blooms. Natural sources of nutrients such as phosphorus or nitrogen compounds can be supplemented by a variety of human activities. For example, in rural areas, agricultural runoff from fields can wash fertilizers into the water. In urban areas, nutrient sources can include treated wastewaters from septic systems and sewage treatment plants, and urban stormwater runoff that carries nonpoint-source pollutants such as lawn fertilizers.

An algal bloom contributes to the natural "aging" process of a lake, and in some lakes can provide important benefits by boosting primary productivity. But in other cases, recurrent or severe blooms can cause dissolved oxygen depletion as the large numbers of dead algae decay. In highly eutrophic (enriched) lakes, algal blooms may lead to anoxia and fish kills during the summer. In terms of human values, the odors and unattractive appearance of algal blooms can detract from the recreational value of reservoirs, lakes, and streams. Repeated blooms may cause property values of lakeside or riverside tracts to decline.

Toxic Blooms

Some algae produce toxic chemicals that pose a threat to fish, other aquatic organisms, wild and domestic animals, and humans. The toxins are released into the water when the algae die and decay. The most common and visible nuisance algae in fresh water, and the species that are often toxic, are the cyanobacteria. A cyanobacterial bloom will form on the surface and can accumulate downwind, forming a thick scum that sometimes resembles paint floating on the water. Because these mats are blown close to shore, humans and wild and domestic animals can come into contact with the unsightly material.

Blooms of toxic species of algae and cyanobacteria can flood the water environment with the biotoxin they produce. When toxic, blooms can cause human illnesses such as gastroenteritis (if the toxin is ingested) and lung irritations (if the toxin becomes aerosolized and hence airborne). Other cyanobacterial toxins are less drastic, and cause skin irritation to people who swim through an algal bloom. Toxicity can sometimes cause severe illness and death to animals that consume the biotoxin-containing water. Cyanobacterial toxins are known to affect bean photosynthesis when they are present in irrigation water. The toxins also can modify zooplankton communities, reduce growth of trout, and interfere with development of fish and amphibians. In some cases, toxins can be bioconcentrated by fresh-water clams.

Microcystins comprise the most common group of about fifty cyanobacterial toxins. Among these toxins are ones that, if ingested in sufficient quantity, can harm the liver (hepatotoxins) or nervous system (neurotoxins). Microcystins can persist in water because they are stable in both hot and cold water. Even boiling the water, which makes the water safe from harmful bacteria, will not destroy microcystins. As a result of this threat, the Canadian government implemented a recommended water-quality guideline of 1.5 μg per litre of microsystin-LR (the most common hepatotoxin), and other countries will likely follow suit. In Canada as well as the United States, there are few reports of injury and no reports of human deaths resulting from microcystins in drinking water, in large part because surface-water sources of drinking water (e.g., reservoirs, lakes, and rivers) must undergo filtration and chlorination at water utilities prior to being distributed to customers. (Cyanobacterial toxins can be removed from water only by activated charcoal filters and chlorination.)

Control Considerations

Repeated episodes of algal blooms can be an indication that a river or lake is being contaminated, or that other aspects of a lake's ecology are out of balance. While cyanobacterial blooms receive the most public and scientific attention, the excessive growth of other algae and other aquatic plants also can cause significant degradation of a lake or pond, particularly in waters receiving sewage or agricultural runoff. Aquatic biologists and other water-quality specialists often are called to identify the causes and recommend management steps to reduce or control the problem.

However, prevention of a problem is always better than trying to fix the problem after it happens. Controlling agricultural, urban, and stormwater runoff; properly maintaining septic systems; and properly managing residential applications of fertilizers are probably the most effective measures that can be taken to help prevent human-induced fresh-water algal blooms.

Marine Ecology

Marine ecology describes the interactions of marine species with their biotic (living) and abiotic (nonliving) environments. The biotic environment includes interactions with other living organisms. The abiotic environment includes aspects of the physical habitat, such as water temperature, chemical composition, depth, and current.

Trophic Levels and Biomass Pyramids

The word "trophic" refers to feeding, and "trophic levels" describe the feeding levels in a food chain. The first trophic level includes species known as primary producers. These organisms produce organic material from inorganic substances using resources from the environment and an external source of energy. Most primary producers are photosynthesizers—that is, they use sunlight as their energy source. Plants and algae are examples of photosynthesizers. Nearly all food chains are based on photosynthesizers, although a few marine food chains depend on bacterial chemosynthesizers, which use a chemical source of energy for production. Chemosynthesizers are discussed later in this entry.

The next level in the food chain, the second trophic level, consists of species that eat the producers. These are sometimes referred to as primary consumers or as herbivores (plant-eaters). The third trophic level consists of secondary consumers, which are also called carnivores (animal-eaters). There can be further, higher trophic levels as well. Finally, there are detritivores and decomposers, both of which feed on dead or decaying organic matter. Much of the decomposition work in food chains is done by bacteria. Species can occupy more than one trophic level, and each trophic level usually has many representatives. Consequently, in most marine ecosystems, trophic interactions are described not by simple chains but as complicated food webs. Many species, including omnivores (eaters of both plants and animals), eat at more than one level of the food web.

Pyramid of Biomass

A pyramid of biomass describes the total amount of biomass, or weight of living matter, that is present in each trophic level. The amount of biomass decreases sharply as one moves up from one trophic level to the next (thus the "pyramid"). That is, the total weight of all the producers in a food chain is greater than the total weight of all the primary consumers, which in turn is greater than that of all secondary consumers. This is because not all the energy that a consumer obtains from food is converted to new biomass.

For example, the consumer must use energy to catch and eat prey. In addition, a lot of energy is lost to metabolism and heat. In fact, only about 10 percent of the energy in one trophic level is passed on to the next level. Because of the decreasing amount of energy available in each trophic level, most ecosystems cannot support more than four or five trophic levels.

Phytoplankton and Zooplankton

In nearly all marine ecosystems, photosynthetic species represent the producers at the base of the food chain.

Photosynthesis requires sunlight, carbon dioxide, and nutrients such as nitrogen and phosphorus, which are found in sea water. Although there are some photosynthetic rooted plants in shallow marine areas, the majority of photosynthetic organisms in the ocean are microscopic algae, or phytoplankton, that drift along with currents in the water. Phytoplankton are found only in the topmost layer of marine water, known as the epipelagic zone, where there is enough sunlight for photosynthesis. Because of the concentration of producers, many consumer species also are found close to the water surface.

The amount of phytoplankton in oceans varies across regions and also changes seasonally. For example, phytoplankton are found in low concentrations in tropical waters, where nutrients are in short supply. Phytoplankton density generally is lowest during the winter, when resources are scarce, and greatest during the spring, when levels of sunlight and nutrients increase. Spring often brings algal blooms, or population explosions of phytoplankton. Algae can be so plentiful during the blooms that they color the water, particularly if they contain red, brown, orange, or purple pigments. The amount of phytoplankton available has implications all the way up the food chain, and blooms in algae populations are often followed by increases in populations of other species.

Phytoplankton are consumed by many organisms, including diverse species of zooplankton. Zooplankton are free-floating consumers and include single-celled protozoa, tiny crustaceans, and the larval stages of species such as mollusks and fish. Zooplankton and phytoplankton are consumed by the nekton, free-swimming marine organisms such as fish, marine mammals, and penguins, as well as species on the ocean bottom, including bivalves, crustaceans, and snails.

Marine Ecosystems

There are numerous types of marine ecosystems. These include coral reefs, tidepools, polar oceans, the abyss, and others.

Coral Reefs

Photosynthetic algae are the producers in coral reefcommunities. The coral reefs represent one of the most diverse marine communities—in fact, a quarter of all marine species are found in or near coral reefs. Photosynthetic algae in coral reefs have a mutualistic relationship with coral, that is, a close association that benefits both members. The algae are sheltered within the corals' calcium carbonate shells and provide nutrients to the corals in exchange. Diverse invertebrates feed on algae and in turn are eaten by reef fish, which are eaten by larger fish species.

Tidal Pools

Tidepool ecosystems are ones that are alternately submerged by water (at high tide) and exposed to air (at low tide). Most of the species found in tidepools are unique to the habitat and are able to survive in both wet and dry conditions. Tidepool food webs, like most marine food webs, are based on algae. Consumer species include bivalves, snails, small fish, and sea anemones. Often the top predators in tidepools are starfish.

Polar Oceans

In polar habitats, such as the oceans around Antarctica and the Arctic, photosynthetic algae also form the basis of the food chain. Algae are eaten by shrimp-like crustaceans called krill, which in turn serve as food for species as diverse as penguins and baleen whales. Penguins are eaten by seals, and penguins and seals are eaten by some whales, particularly killer whales, which are the topmost predators in the system.

Kelp Forests

Kelp forests are marine ecosystems characterized by gigantic species of floating photosynthetic algae called kelp. Kelp can reach lengths of up to 80 meters (about 262 feet). A number of crustaceans feed on kelp, as do sea urchins. These are eaten by larger organisms, such as sea otters.

The Abyss

The deepest part of the ocean, the abyss, extends to depths as great as 6,000 meters (about 19,685 feet, or 3.7 miles). These deep-sea environments are characterized by cold temperatures and lack of light. Consequently, no photosynthesizers exist, although a diverse array of detritivores feed on dead organic matter that floats down from above. There are also numerous predatory and parasitic species.

Hydrothermal Vents

Hydrothermal vents are cracks or openings in the ocean floor where hydrogen sulfide, metals in solution, and other chemical compounds escape into the sea water. Certain specialized chemosynthetic bacteria live in these hot areas and produce organic matter from hydrogen sulfide. They form the base of unusual food webs in this specialized habitat.

Chemosynthetic bacteria are eaten by specialized vent worms, clams, and mussels, which in turn provide food for octopuses and other species. The hydrothermal vent crab is at the top of the food chain in vent environments. These unique deep-sea ecosystems are more diverse than most deep-sea environments.

Bioconcentration

A feature of the ordered structure of food chains is that substances, especially pollutants, become concentrated in large amounts at higher trophic levels. This process is called bioconcentration. Bioconcentration occurs because species that consume pollutants do not excrete them but, rather, store them in bodily tissues, where they accumulate over time. For organisms high in the food chain, their prey organisms have already concentrated pollutants from multiple prey of their own, and so forth. This results in high concentrations of pollutants in species that eat high in the food chain. An issue of particular relevance in aquatic ecosystems is mercury poisoning. Mercury causes severe health problems in human and other species, including brain damage, and is bioconcentrated in the upper trophic levels. This is why people, pregnant women in particular, are often advised against eating predatory fish such as tuna and swordfish.

LIFE IN EXTREME WATER ENVIRONMENTS

Beginning in the early 1990s, scientific knowledge of the environmental limits of microbial life on earth expanded dramatically as microbiologists applied new methods of molecular biology over a broad range of environmental extremes. Microbial species are now known to occupy a vast range of environments that previously were unimagined. New discoveries have revolutionized scientific understanding of earth's biosphere, opened up new views of the history of terrestrial (land-based) life, and increased the possibilities that life could develop elsewhere in the cosmos.

The name applied to this new research area of biology is extremophiles research. Extremophiles (literally "extreme-loving") are defined as organisms that occupy environments judged by human standards as harsh. These encompass both physical and chemical extremes. Examples of water environments characterized as extreme are summarized in the accompanying table.

Different classes of extremophiles have been defined based on the nature of the environments where they are found. For example, extremophiles that have adapted to high temperatures are called thermophiles. Those that require cold temperatures for growth and reproduction are called psychrophiles (as opposed to other organisms that can tolerate occasional cold temperatures and are not considered extremophiles). Those that love acidic environmentse, with low pH) are called acidophiles, whereas those found in highly alkaline conditions (high pH) are alkaliphiles. Organisms that live under high pressure are called piezophiles, and those found in high-radiation environments are as yet unnamed. Some organisms occupy more than one environmental extreme simultaneously, and are known as polyextremophiles. An example is the archaebacterial species, Sulfolobus acidocalderius, which thrives in boiling mudpots at temperatures exceeding 80°C (176°F) and at acidities less than pH 3. Although mostly microbial, extremophiles include a few species of multicellular organisms such as worms, amphibians, mollusks, and crustaceans.

Physical Extremes

Temperature

Microorganisms are now known to thrive over a broad range of physical extremes in temperature. For high temperatures, this environment includes geysers and hot springs, boiling mudpots, and hydrothermal vents on the deep seafloor. In the latter case, where vent temperature can reach 400°C (752°F), the high hydrostatic pressure prevents vent water from boiling, and thermophilic species exhibit growth up to a temperature of about 114°C (237°F). (Under atmospheric pressure, water boils at 100°C, or 212°F.)

At the lower end of the temperature environment is sea ice, ground ice, permafrost (icy soils), and subglacial lake environments like Lake Vostok in Antarctica, a deep subglacial lake located more than 4 kilometers (2.5 miles) beneath the Antarctic ice sheet. In salt-water environments, water can remain liquid below 0°C (32°F) because dissolved salts lower its freezing point. Some psychrophilic species, such as those found in brine films, are known to be active down to -15°C (5°F). Some complex multicellular organisms, like the wood frog, can tolerate the freezing of up to 65 percent of their body water during winter hibernation.

Pressure

Pressure, which is measured relative to atmospheric pressure at sea level (where 1 bar roughly equals 14.5 pounds per square inch), increases with depth in the oceans. In the ocean, this hydrostatic pressure goes up at the rate of approximately 1 bar per 100 meters. Measured within the crust, lithospheric pressure increases at a rate almost twice hydrostatic. Live microorganisms obtained from the Mariana Trench, the deepest place in the oceans (10.9 kilometers, or 6.8 miles), have been successfully grown under surface conditions, whereas others have been shown to be obligate piezophiles that grow only at high pressure.

Pressure decreases with increasing altitude, such that at 10 kilometers (6.2 miles) above the Earth's surface, the pressure is only about one-fourth that at sea level. Organisms have been discovered growing on the top of Mount Everest, the highest point on the Earth's surface (more than 8.8 kilometers [5.4 miles]). Viable spores of bacteria and fungae have even been collected from the lower stratosphere.

Radiation

Radiation is energy that travels as either particles (e.g., highenergy neutrons, protons, electrons, or ions) or waves (e.g., X-rays, gamma rays, or ultraviolet rays). The bacterium Deinococcus radiodurens, which has been found growing on the fuel rods of nuclear reactors, is a famous example of an extremophile that can tolerate high levels of radiation.

Chemical Extremes

pH

Chemical extremes in the environment include pH, which ranges from values of less than 0 (extremely acidic) to more than 14 (extremely alkaline or basic). In nature, microorganisms have been shown to occupy nearly the entire range of pH. Some species of bacteria have been found living in a acid mine drainage at a pH of approximately 0.5. Others live in soda lakes, such as those found in the western United States and Egypt, where the highly alkaline waters can reach a pH of 11.

Salinity

Life also occupies an equally broad range of salinity. Salt-loving halophiles live in salt plains, evaporation ponds at saltworks, and natural salt lakes (e.g., the Dead Sea, Israel and the Great Salt Lake, Utah). Halophiles also live within hypersaline brines that exist around deep-sea vents and in deep subsurface rock formations. In nature, salinities can range from fresh water, with very low salinity, to super-saturated brines. At very high concentrations, salt precipitates, often entrapping microorganisms.

Desiccation

The ability to survive desiccation (extreme drying) has been demonstrated for both vegetative cells and reproductive spores of many microbial species. In the driest deserts on Earth, microbial species (so-called "endoliths") often survive by living inside porous rocks where they are protected from ultraviolet radiation. The coldest desert environments on Earth are found in the dry valleys of Antarctica. These polar deserts harbor many types of endolithic communities dominated by cyanobacteria, algae, and fungi. Antarctic endoliths live just a few millimeters beneath rock surfaces in limestones or in translucent, quartz-rich sandstones. Decades may pass with no rain, but when it comes, these organisms spring to life, using the available light, water, and nutrients to quickly grow and reproduce before drying out and again becoming dormant (inactive).

Aphotic and Anoxic Environments

Even though photosynthesis accounts for more than 99 percent of the energy that powers the biosphere, thermal and chemical energy sources within the Earth can provide forms of energy capable of supporting complex ecosystems. Consequently, extremophiles can also be found in aphotic (non-light) environments, such as deep in the ocean or in the Earth's subsurface. In hydrothermal vent environments on the ocean floor, complex ecosystems have been found in which the organisms (including large, multicellular animals) derive their energy entirely from chemical sources provided by the hot fluids issuing from the vent. Single-celled forms of life also survive and grow in the deep subsurface of Earth, within the tiny pore spaces and fractures of endurated rock. These are aphotic environments where sunlight does not penetrate; consequently, organisms living there must use chemical energy sources for their metabolism. Some subsurface microbes do depend on photosynthetically-derived organic matter that washes down from the surface, but many so-called lithoautotrophic species (which literally means "self-feeding from rocks") use simple byproducts of chemical weathering of rocks to extract energy from the environment. For example, oxidation reactions associated with the weathering of basalts in an oxygen-free environment may lead to the release of hydrogen. The hydrogen released is used by methanogens to produce methane and energy. Although some scientists have questioned the evidence for hydrogenbased microbial ecosystems in deep basalt formations, the possibility of an active microbial community at great depths could have implications for subsurface storage of highly radioactive materials and other wastes. Microbial interactions could act to weaken containers, leading to leakage and the undesirable spreading of waste materials.

Methanogens are microbes that live in anoxic (non-oxygen) environments, which can include some swamps, rice paddies, or certain highly enriched lakes, ponds, or streams, and their sediments. Methanogens combine carbon dioxide (CO_2) and hydrogen (H_2) to produce organic matter, while releasing methane gas as a byproduct. Wetlands and rice paddies (agricultural wetlands) account for half the total methane produced globally. Methane is a greenhouse gas, and its rate of increase in the atmosphere is exceeding that of CO_2. Human activity has played a major role in this methane increase.

Implications and Benefits

The ability of some extremophiles to survive harsh conditions similar to those found on other planets has raised the possibility that life might exist beyond Earth. As an example of this survival ability, halophiles have been cultivated from inclusions of brine contained in salt crystals deposited hundreds of millions of years ago. Microbes also have been isolated from Siberian permafrost, where they have remained in deep freeze for more than three million years. Equally impressive are bacteria germinated from spores preserved in Dominican amber dated at more than 30 million years old. Given the propensity for prolonged survival in these types of environments, could an extraterrestrial biota someday be discovered within brines, salts, or ices on another planet, like Mars or a moon like Europa?

Cellular enzymes extracted from extremophiles have spawned a multibillion dollar biotechnology industry. The enzymes are used in industrial and medical applications, ranging from the production of stone-washed jeans, to creating artificial sweeteners, to genetic fingerprinting. One thermophile that lives in hot springs is the source of the heat-stable deoxyribonucleic acid (DNA) polymerase enzyme used in polymerase chain reaction (PCR). PCR forms part of the foundation of much of the biotechnology industry. Proteins produced by psychrophilic organisms may one day prove useful in coldfood preparation and in detergents for washing in cold water.

Microscopic Life in Water

Introduction

There are many microscopic plants and animals that can be found in most lakes, ponds, swamps, rivers, streams, and even puddles. These tiny life forms are called plankton. Plant plankton is called phytoplankton, and animal plankton is called zooplankton. In this activity you will be examining common types of zooplankton. A small sample of water, such as the few drops you can collect in an eyedropper, usually will not contain many zooplankton. But ten gallons of water will. Since collecting and examining ten gallons of water could take a very long time, scientists use something called a plankton net. Plankton nets collect all the zoo-plankton from a large amount of water and concentrates them into a smaller amount of water. From this smaller amount of water you can collect a lot of zooplankton using an eyedropper.

Challenge

Make a plankton net and use it to collect animals swimming in the water. Observe these animals at varying powers of magnification.

Tools and Materials

- coat hanger;
- leg from a pair of old nylon panty hose ;
- needle;
- sewing thread;
- scissors;
- small plastic bottle;
- eyedropper;
- hinged well slide;
- rubber band;
- hand lens;
- pencil;

- eraser;
- paper;
- microscope;
- Instructions.

Bend a coat hanger to make a circle with a handle. Cut off a leg from a pair of old nylon panty hose. Sew the thigh-end of the nylon onto the coat hanger by folding about an inch over the coat hanger and sewing. Now cut off the foot. Using a rubber band, attach a small plastic bottle (such as an empty vitamin bottle) onto the foot-end of the nylon by wrapping the rubber band around the top of the bottle.

Slowly drag the net through the water. The water will enter the net and leave the net through the nylon sides. The animals will enter the net and collect in the plastic bottle. Carefully remove the rubber band from the bottle to free it from the net. Now use the eyedropper to transfer some of the animals onto the well slide. Begin by observing these animals with your hand lens. If you can see any animals at 5x, draw them. Then examine the water sample at 10x and draw what you see. Now you can use a microscope. Begin at 75x and draw what you see. Increase the magnification to 150x and make a new drawing. Finally, increase magnification to 300x and create a final drawing. Don't forget to write the magnification on each of your drawings.

Further Exploration

Collect some filamentous, or "hairy" algae, from a pond. This is a good place to find zooplankton because many of the animals live on the algae, either eating the algae itself or other animals that live in the algae. Place it in a jar with pond water. Don't let the algae dry out because it will kill the animals. Cut off a small amount of algae with a pair of scissors and place it in the well slide with a drop of water. Look at your sample, first with the hand lens, then with a microscope.

7

Importance of Water

INTRODUCTION

Water is an essential element for life. Many people must confront daily the situation of an inadequate supply of safe water and the very serious resulting consequences. The intention of this paper is to present some of the human, social, economic, ethical and religious factors surrounding the issue of water.

The Holy See offers these reflections on some of the key issues in the agenda of the 3rd World Water Forum (Kyoto, 16th-23rd March 2003), in order to contribute its voice to the call for action to correct the dramatic situation concerning water. The human being is the centre of the concern expressed in this chapter and the focus of its considerations.

The management of water and sanitation must address the needs of all, and particularly of persons living in poverty. Inadequate access to safe drinking water affects the well being of over one billion persons and more than twice that number have no adequate sanitation. This all too often is the cause of disease, unnecessary suffering, conflicts, poverty and even death. This situation is characterized by countless unacceptable injustices.

A FAR-REACHING QUESTION

Water plays a central and critical role in all aspects of life—in the national environment, in our economies, in food security, in production, in politics. Water has indeed a special significance for the great religions. The inadequacy in the supply and access to water has only recently taken centre stage in global reflection as a serious and threatening phenomenon. Communities and individuals can exist even for substantial periods without many essential goods. The human being, however, can survive only a few days without clean, safe drinking water.

Many people living in poverty, particularly in the developing countries, daily face enormous hardship because water supplies are neither sufficient nor safe. Women bear a disproportionate hardship. For water users living in poverty this is rapidly becoming an issue crucial for life and, in the broad sense of the concept, a right to life issue. Water is a major factor in each of the three pillars of sustainable development—economic, social and environmental. In this framework, it is understood that water must meet the needs of the present population and those of future generations of all societies. This is not solely in the economic realm but in the sphere of integral human development. Water policy, to be sustainable, must promote the good of every person and of the whole person.

Water has a central place in the practices and beliefs of many religions of the world. This significance manifests itself differently in various religions and beliefs. Yet two particular qualities of water underlie its central place in religions: water is a primary building block of life, a creative force; water cleanses by washing away impurities, purifying objects for ritual use as well as making a person clean, externally and spiritually, ready to come into the presence of the focus of worship.

THE WATER ISSUE: SOME ETHICAL CONSIDERATIOINS

The principle water difficulty today is not one of absolute scarcity, but rather of distribution1 and resources. Access and deprivation underlie most water decisions. Hence linkages between water policy and ethics increasingly emerge throughout the world.

Respect for life and the dignity of the human person must be the ultimate guiding norm for all development policy, including environmental policy. While never overlooking the need to protect our eco-systems, it is the critical or basic needs of humanity that must be operative in an appropriate prioritisation of water access. Powerful international interests, public and private, must adapt their agendas to serve human needs rather than dominate them. The human person must be the central point of convergence of all issues pertaining to development, the environment and water. The centrality of the human person must thus be foremost in any consideration of the issues of water. The first priority of every country and the international community for sustainable water policy should be to provide access to safe water to those who are deprived of such access at present.

The earth and all that it contains are for the use of every human being and all peoples. This principle of the universal destination of the goods of creation confirms that people and countries, including future generations, have the right to fundamental access to those goods which are necessary for their development.

Water is such a common good of humankind. This is the basis for cooperation toward a water policy that gives priority to persons living in poverty and those living in areas endowed with fewer resources. The few, with the means to control, cannot destroy or exhaust this resource, which is destined for the use of all. People must become the "active subjects" of safe water policies. It is their creativity and capacity for innovation that makes people the driving force toward finding new solutions. It is the human being who has the ability to perceive the needs of others and satisfy them.

Water management should be based on a participatory approach, involving users, planners and policy makers at all levels. Both men and women should be involved and have equal voice in managing water resources and sharing of the benefits that come from sustainable water use. In a globalized world the water concerns of the poor become the concerns of all in a prospective of solidarity. This solidarity is a firm and persevering determination to commit oneself to the common good, to the good of all and of each individual. It presupposes the effort for a more just social order and requires a preferential attention to the situation of the poor. The same duty of solidarity that rests on individuals exists also for nations: advanced nations have a very heavy obligation to help the developing people. The principle of subsidiarity acknowledges that decisions and management responsibilities pertaining to water should take place at the lowest appropriate level. While the water issue is global in scope, it is at the local level where decisive action can best be taken. The engagement of communities at the grassroots level is key to the success of water programmes.

While vital to humanity, water has a strong social content. It is highly charged with symbolism and is one of the essentials of life. Among the important social characters of water is its role in human nourishment, health and sanitation as well as peace and conflict avoidance.

Water for Food and Rural Development

Agriculture represents a key sector in the economies of developing countries and cannot be sustained without sufficient water. In most of these countries agricultural activities are a major source of livelihood and an essential dimension of local social cohesion and culture. This activity is carried on by small farmers in rural areas, very often with huge constraints. However, it must be remembered that, in the end, the dominant use of water around the world will continue to be water for food security.

People living in rural areas, many times in poverty, can be driven by necessity to exploit beyond sustainable limits the little land they have at their disposal. Special training aimed at teaching them how to harmonise the cultivation of land with respect for water and other environmental needs should be encouraged. Where possible, cooperative efforts of water management and use should be encouraged.

Participation suffers when large portions of a population lack skills and knowledge to engage in the issue before them. It should not be overlooked, however, that often those lacking formal education possess traditional forms of knowledge that can be vital and decisive in addressing and solving the question of water. The special knowledge of indigenous people should be esteemed.

In the context of rural development, a shift is needed, however, in the emphasis from the traditional irrigation to other means that focus on the needs of the poor and their food insecurity. The challenges are to develop water-saving technologies and to structure incentives to encourage development.

Lands that have been damaged by waterlogging and salinization must be reclaimed through drainage programmes. New irrigation development needs to be carried out with proper environmental impact assessment. Policies must be encouraged that develop sustainable irrigation and harness the wider potential of rainfed farming, incorporating water management for gardens and foods from common property resources.

Safe Drinking Water, Health and Sanitation

Three crucial concerns are present in the relationship between water and health: managing quantity constraints faced by water-poor countries and their impact on human activities; the maintenance of water quality in the face of growing demand; and the direct link between health and water as pertains to diseases.

Management of water quantity can be carried out by revising the allocation of water to different users. Better maintenance and repair of existing water systems can often significantly increase the water supply. Water conservation methods such as rainwater harvesting, fog condensation and underground dams should be studied for use where appropriate along with stabilization ponds for wastewater and treatment technology for the use of wastewater for irrigation.

Water shortages can be substantially overcome through further development and use of treated urban wastewater for use in agriculture. This has considerable potential and if carefully managed carries only very limited risks and associated difficulties. The problem of maintaining and improving water quality is especially acute in the more urbanized areas, predominantly in developing countries. This is most often hampered by a failure to enforce pollution controls at the main point source and the inadequacy of sanitation systems and of garbage collection and disposal.

Most of the diseases that contaminate water come from animal or human waste and are communicable. These diseases have health effects that are heavily concentrated in the developing world, and within that context particularly among poor urban populations. Wastewater is often the medium through which these can affect humans. Whether it relates to quantity, quality or disease, the trend away from centralized government agencies and towards empowering local governments and local communities to manage water supplies must be emphasised. This necessitates building community capacities, especially in the area of personnel, and the allocation of resources to the local level.

Peace and/or Conflict

Growing pressure due to increasing demand for water can be a source of conflict. When water is scarce, competition for limited supplies has lead nations to see water as a matter of national or regional security. History provides ample evidence of competition and disputes over shared fresh water resources.

Identifying potential trouble areas does little good if there are no effective and recognized mechanisms for mitigating tensions. Existing international water law may be unable to handle the strains of ongoing and future problems. But some mechanisms for reducing the risks of such conflicts do in fact exist. These need renewed international support and should be applied more effectively and at an earlier stage of potential conflicts.

At the international level, conflicts tend to focus on shared river basins and transboundary waters, especially when combined with circumstances of low water availability. Tensions arise with increasing frequency over projects to dam or divert water by countries in a powerful position upstream from their neighbouring countries. Water has always been acknowledged for its role in production and thus in the economy. However, in recent years increased emphasis has been given to the economic value of water.

THE ECONOMICS OF WATER

The economics of water is one of the most important aspects of water resource management that needs to be balanced with cultural and social concerns. The concept of treating water as an economic good is valid but the practice of doing so can be challenging. The use of water for industry and energy are of great importance in terms of the amounts of water used, the cost of investments to provide the water and the economic significance of the resultant production. Every water policy must address the underlying economic issues.

The aim of treating water as an economic good should be to accord water its proper economic value and enable the water economy of the country to be integrated with the broader national economy. Policies relating to the economics of water should ensure optimum efficiency and the most beneficial use while meeting the required objects of social development and environmental sustainability. There are increasing instances, however, of the commercialisation of water and water services. The most delicate and sensitive point in the consideration of water as an economic good is to ensure that a balance is maintained between ensuring that water for basic human needs is available to the poor and that, where it is used for production or other beneficial use, it is properly and appropriately valued.

Water and Energy

Hydroelectric power is an important source of clean energy. It provides approximately twenty percent of total electricity production worldwide and brings notable economic and environmental benefits. For poor mountainous regions it offers one of the few avenues for economic growth via electricity exports. However, too often in the past such projects have been accompanied by devastating environmental costs. Policy discussion in this area has been dominated by big dams to the neglect of issues such as small-scale hydropower and water use for cooling in thermal power plants. While most of this water re-enters the water system, the significant change in temperature and in some cases quality, has serious environmental and resource implications. Dams still remain today one the most contentious development issues for the water sector.

Private Sector Engagement and Privatisation

Water by its very nature cannot be treated as a mere commodity among other commodities. Catholic social thought has always stressed that the defence and preservation of certain common goods, such as the natural and human environments, cannot be safeguarded simply by market forces, since they touch on fundamental human needs which escape market logic. Water has traditionally been a State responsibility in most countries and viewed as a public good. Governments worldwide, for diverse political and social considerations, may indeed often provide large subsidies to insulate water users from the true cost of water provision. Being at the service of its citizens, the State is the steward of the people's resources which it must administer with a view to the common good. At the same time, in the interest of achieving more efficient sustainable water services, private sector involvement in water management is growing. It has however proved to be extremely difficult to establish the right balance of public-private partnerships and serious errors have been committed. At times individual enterprises attained almost monopoly powers over public goods. A prerequisite for effective privatisation is that it be set within a clear legislative framework which allows government to ensure that private interventions do in actual fact protect the public interest.

The debate today is not whether the private sector will be involved but how and to what extent it will be present as the actual provider of water services. In any formation of private sector involvement with the state, there must exist a general parity among the parties allowing for informed decisions and sound agreements. A core concern in private sector involvement in the water sector is to ensure that efforts to achieve a water service that is efficient and reliable do not cause undue negative effects for the poor and low-income families. The debate surrounding water has historically been largely confined to socio-economic issues.

Today, in the context of sustainable management of water resources, the environmental aspect is coming to the forefront along with water's role in supporting ecosystem functioning and species. This approach to water resources has focussed on sustainable use and on ensuring water utilization that is environmentally sound. A specific proposal to protect aquatic ecosystems and fresh water living resources has been put forward over the years reflecting the extreme threats that exist for many wetlands, rivers and lake ecosystems, deltas and other areas. Systematic changes to policy approaches are now needed, moving away from a traditional supply-side technical focus to one in which environmental issues are seen as integral to water policies and practices.

Policy goals and·priorities have in some cases to be re-ordered with frequent use of Environmental Impact Assessments as determinants of decisions on water investments. There is, however, a lack of adequate human resources in this sector. This calls for planning and investments in human resource development.

Environmentally Sound Sanitation

Conventional forms of centralised sanitation are coming under increasing criticism due to huge operating and maintenance costs but more importantly their high water consumption and the groundwater pollution that can result. Further these types of wastewater and sewage disposal systems usually deprive agriculture, and consequently food production, of valuable nutrients. An alternative approach towards ecologically and environmentally sound sanitation is offered by a concept referred to as "ecological sanitation".

This takes the principle of environmental sanitation further in that their focus is keeping the environment clean and safe and preventing pollution. It includes wastewater treatment and disposal and disease prevention activities. It is an approach premised on recycling principles with a key objective of promoting a new philosophy of dealing with what has been regarded as waste.

Disaster Mitigation and Risk Management

A people centred pro-poor policy on water management must address the question of water related hazards such as floods, droughts, desertification, tropical storms, erosion and various kinds of pollution. Many so-called natural disasters are in fact man made in their roots, due to inadequate attention to the environment and the consequences of human actions or indeed inaction. Once again, it is the poor who suffer most when they are exposed to such dangers. But everyone's security is at risk. More can be done in the areas of monitoring and forecasting of extreme events especially through more efficient early warning system and technical cooperation between poor and more developed countries in devising planning strategies and setting up appropriate infrastructures.

Climate variability and change are now recognized as being an essential dimension of such evaluation. Efforts of humanitarian assistance in response to disasters relating to water must identify the faults which gave rise to such occurrences and ensure that they do not recur. Post disaster reconstruction is not a question of reconstructing the past, but of building for a safer and more ecologically sustainable future.

The water that exists today would be enough to meet human needs if it were equitably distributed throughout the world. Since it is not, there arise situations of scarcity; some due to natural causes and others due to a range of human activities.

Population

World population has continued to grow throughout history. While the human demand for fresh water has risen steadily, since 1940 the global water withdrawals have risen even faster than the rate of population growth. It is correct to deduce that more people need more water. However, to attribute to population growth a disproportional role misrepresents the true picture. The principal cause in increased demand is not in itself the mere growth of population but the disproportionate and unsustainable use of water for production and consumption by populations in developed countries. The ever growing concentration of a very high percentage of the world's population in large urban areas, especially in mega-cities, is going to propose new challenges for water and sanitation management, which will seriously impact the short-term and long-term local demand for water.

Politics

Water is a political issue. There is little today that cannot be achieved technically. What is needed is political effectiveness, political will and effective governance. The political arena is where decisions of water utilisation will take place. The solution to water problems requires the interaction of many spheres and sectors. This interaction must take account of the objectives of safe drinking water, sanitation and food security for all. Politics must ensure proper interaction, through setting correct priorities and the equitable allocation of resources, as well as through fostering interaction between institutions and the engagement and support of local communities, who are the most directly affected. Political will and effective follow through is required for successful action in the water sector. The long-term viability of a country's water supply infrastructure depends on leadership and vision of political leaders, at national and local levels and their capacity to get things done.

New legislation and institutional changes will be needed in many countries to form the framework within which the politics of water supply can be realised. A larger portion of the national budget may need to be directed to the water sector. Political leaders are crucial in generating genuine political support and vision in order to provide the motivation for such changes. Often the institutional structure of the water sector at government level and the water portfolio is moved about between different ministries and many times is the result of political uncertainty and a lack of political responsibility. The international political arena must be given its proper role in seeking and formulating global strategies to address water issues. The issue of water cuts across so many areas relating to sustainable development and poses considerable challenges to politics at the international level. Action-orientated responses to the challenges is what the people of the world await.

A Right to Water

A major achievement of recent history has been the ability to elaborate, within the framework of the United Nations, a network of international instruments formally identifying and proclaiming a broad spectrum of universally recognized human rights. Although access to water is a precondition to many of these rights, "clean drinking water" is explicitly mentioned only in the Convention on the Rights of the Child. It is however to be found in some regional human rights documents and national Constitutions.

Sufficient and safe drinking water is a precondition for the realization of other human rights. It is argued that water was so fundamental a resource that, just as a right to air was not identified, water was not explicitly mentioned at the time the fundamental human rights documents were drawn up but was understood as a given which the drafters implicitly included. Furthermore, several of the explicit rights protected by conventions and agreements, such as rights to food, clothing, housing and medical care and necessary social services, cannot be attained or guaranteed without also guaranteeing access to clean water.

There is a growing movement to formally adopt a human right to water. The dignity of the human person mandates its acknowledgement, along with the sound and logical argumentation found in the concept of implicit inclusion. Water is an essential commodity for life. Without water life is threatened, with the result being death. The right to water is thus an inalienable right.

The challenge remains as to how such a right to water would be realized and enforced at the local, national and international levels. Just as, for example, the acknowledgement of the right to food has not eliminated hunger, the promotion of the right to water is a first step and needs careful implementation thought to arrive to the desired goal of access to safe drinking water for all. A right to adequate and safe drinking water should be interpreted in a manner fully consistent with human dignity and not in a narrow way, by mere reference to volumetric quantities and technologies or by viewing water primarily as an economic good.

Poverty

Poverty is the most important factor related to the sustainable provision of basic water and sanitation services. The unavailability of basic services is a primary measure of poverty and poverty is the primary obstacle in the effective provision of basis services. Water scarcity has more dramatic effects for the poor than for the wealthy. The cost of even minimal basic water services is so high that the poor may never be able to afford them.

Sustainable water policies will not be attained in areas which are impoverished in many other aspects. Poor services are a symptom of something fundamental. Authorities are unable to provide the institutional framework and the infrastructures to regulate the sector. Development at the institutional level is needed whereby the priority of water is clearly identified. The authority and responsibility to enable services to operate efficiently must be provided. This will require structures for environmental and economic regulation. The water services in many developing countries are however still plainly inadequate in providing safe water supplies. This situation is so dramatic that it will not be overcome without increased development assistance and focused private investment from abroad. Funds released through debt relief could well be utilized in improving water services. Country partnerships can provide a method of institutional building and reform whereby a long-term link can be formed between the water sector of a developed country and that of a developing country. International poverty reduction strategies should focus explicitly on the water needs of the poorest populations.

National and local financial support for the water sector must also increase. Where subsidies are necessary, and they will be necessary, they should carefully target poor and families living in poverty rather than being applied generally. Following consultation at the community level, policies on water and related public health and environmental sectors need to be revised and where lacking established. After such policy change there is need to create or revise the body of laws impacting water that will effectively obtain and allocate the necessary supply of it. Poverty is about people and their ability to realize their God-given potential. The poor show extraordinary creativity in seeking means of survival in the absence of adequate services. This creativity is a resource which should not be overlooked in working together to build up sustainable communities and avoid the creation of dependence.

CONCLUSION

Water is an essential element for life. Right throughout human history water has been looked on as something intertwined with humankind. Human beings live alongside water and are nourished by water. It is a source of beauty, wonder and relaxation and refreshment. Our very contact with nature has a deep restorative power. It is no accident that people chose places associated with water for the holidays, in order to renew and regenerate themselves. Water has an aesthetic value.

In the Judeo-Christian Holy Book, God is presented as the source of living water beside which the just man can find life. Because the Bible was written in a part of the world where water is scarce, it is not surprising that water features significantly in the lives of the people. Due to the scarceness of water in the lands of the Scripture, rainfall and an abundance of water was seen as a sign of God's favour and goodness.

Water is a primary building block of life. Without water there is no life, yet water, despite its creative role, can destroy. The Bible opens precisely with the image of the divine spirit hovering over the water at the creation of the universe. In the accounts of creation contained in the first two chapters of the Bible, it is from the midst of the waters that dry land is made to appear, while living reptiles and rich life forms are made to swarm the waters. It is also water that moistens the earth for other forms of life to appear.

The separation of the elements permits them to interact in a positive sense, recognizing the intrinsic value of each. Disorder and confusion among the elements provokes a return to the primeval chaos. Humankind is thus called to live in harmony with creation and to respect its integrity.

Conservation of water is good because it provides for future generations that fundamental good which nourishes and allows us to protect such a source of power beauty and many other nice things.

None of the issues presented here is done in isolation. Only in a true holistic approach can the human being confront the challenges set forward in addressing the issue of water. The Holy See's contribution is presented with the conviction of the central role of the human being in caring for the environment and its constitutive elements. Only when humankind respects the integrity of creation, in conformity to God's providential plan, will we reach a true appreciation of the significance of water in creation and for humankind.

8

Controlling Bacterioplankton

INTRODUCTION

Three is evidence that the potentially harmful solar ultraviolet-B (UV-B, 280-320 mm) radiation penetrates much deeper into the ocean's water column than previously thought UV-B radiation is also responsible for photochemical degradation of refractory macromolecules into biologically labile organic compounds . It thus seems reasonable to assume that UV-B radiation might influence the cycling of organic matter in the sea, which is believed to be largely mediated by bacterioplankon Here we report that bacterioplankton activity in the surface layers of the oceans is suppressed by solar radiation by about 40% in the top 5 m of the water column in nearshore waters, whereas in oligotrophic open oceans suppression might be detectable to a depth of >10m. Bacterioplankton from near-surface (0.5 m depth) waters of a highly stratified water column were as sensitive to surface UV-B radiation as subpycnocline bacteria, indicating no adaptative mechanisms against surface solar radiation in near-surface bacterioplankton consortia. Surface solar radiation levels also photochemically degrade bacterial extracellular enzymes. Thus elevated UV-B radiation due to the destruction of the stratospheric ozone layer might lead to reduced bacterial activity and accompanying increased concentration of labile dissolved organic matter in the surface layers of the ocean as bacterial uptake of this is retarded.

In evaluating the possible effects of natural solar radiation, we have to distinguish possible effects of short- and long-term exposure on the bacterial community. In a mixed water column, short-term exposure should predominate. We determined the influence of short-term (30 min) solar radiation on thymidine and leucine incorporation into bacterioplankton of the surface layer of the sea by incubating surface water at different UV-B radiation levels. Surface water samples were collected with acid-cleaned Niskin bottles in the northern Adriatic Sea from depths ranging from 0.5 m to 2 m. Bacterial cell production as well as bacterial protein production was affected by UV-B radiation typically for surface layers Compared with bacterial activity in the dark, a 50% reduction in thymidine incorporation was observed at UV-B radiation levels of 1.3 W m^{-2}, whereas a 50% reduction in leucine incorporation was detectable at a radiation of about 0.6 W m^{-2}. Thus we conclude that, at least during short-term exposure, bacterial biomass production is more affected than cell division and hence cell production. The higher bacterial activity in the low UV-B radiation range (<0.2 W m^{-2}) compared to the dark activity might be caused by the release of easily metabolizable substances from the photosynthetic activity of phytoplankton.

In a stratified water column mixing is restricted and bacterioplankton in the surface waters may be exposed to high levels of UV-B during their entire life cycle, whereas bacterioplankton from the subpycnocline waters never experience noticeable levels of ultraviolet radiation. We addressed this problem by exposing surface water from a highly stratified water column to surface solar radiation levels on cloudless days for 4 h. Before and immediately after exposure to natural radiation levels, thymidine incorporation into bacterial DNA was measured as well as 4 and 8 h after exposure to solar radiation . After 4 h of exposure to surface light regimes, about 48% of suppression in bacterial growth is caused by UV-B and ~40% of the observed suppression is caused by photosynthetic active radiation (PAR) plus UV-A. After 4 h in dark, bacterial production in the bottle kept continuously in the dark and in the quartz bottle covered by a Mylar foil (PAR + UV-A) reached almost identical rates, indicating rapid recovery from light exposure in the absence of UV-B. In the incubation flasks exposed additionally to UV-B radiation, thymidine incorporation was suppressed by about 56% even after 4 h in darkness, indicating prolonged suppression of bacterial growth after UV-B exposure.

To concentrate on possible adaptations to high UV-B radiation levels, we incubated water from the subpycnocline layer (20 m depth) and compared the rate of thymidine incorporation into bacterial cells under surface water light regimes (PAR+UV-A+UV-B) with thymidine incorporation rates of surface layer bacteria. After exposure for 4 h, no significant difference was detectable in thymidine incorporation rates between subpycnocline waters and surface waters; we therefore conclude that surface water bacterioplankton does not exhibit any adaptations to UV-B radiation despite the fact that they are exposed daily to a certain level of UV-B.

Additionally, bacterial extracellular enzymatic activity was measured using fluorogenic substrate analougus . After exposure to surface solar radiation levels for 4 h, bacterial extracellular enzymatic activity was reduced to less than one third of the value measured before exposure, whereas dissolved enzymatic activity in particle-free water was reduced to 50-60% of the values before UV-B exposure, depending on the enzymes tested These results indicate that enzyme expression is only repressed to a minor extent; more importantly, the enzymes are photolytically cleaved during UV-B exposure . Therefore UV-B radiation suppresses bacterial activity and concurrently photolytically cleaves macromolecules and thereby substitutes bacterial enzymatic activity in cleaving dissolved organic matter (DOM).

BACTERIAL METABOLISM

Bacterial metabolism is closely coupled to primary production. Consequently, bacterial activity is highest in the upper mixed layer of the ocean where UV-B also penetrates at radiation levels high enough to influence bacterial growth. Extrapolating our measurements on UV-B penetration, we calculate that UV-B influences bacterial activity down to a depth of 10 m at around noon. Furthermore, the 1% UV-B radiation level in tropical waters is at about 30 m depth at around noon. An inhibition of bacterial activity by about 30 to 40% (depending on whether thymidine or leucine incorporation techniques have been used) is noticeable at a UV-B radiation of 0.4 W m^{-2}; this radiation has been measured between 10:00 and 14:00 off Belize on a cloudless day down to a depth of 3.5 m. This finding has several implications in evaluating the role of bacteria in the sea. Because even surface water bacterioplankton communities from highly stratified water columns do not have adaptative capacities to protect the cells from harmful UV-B, suppression of bacterial growth due to UV-B penetration in the surface layer of the ocean should be considerable. The lack of pigmentation in bacterioplankton is surprising as phytoplankton species are able to synthesize pigments to protect the cells from harmful radiation.

The increase in global UV-B radiation due to ozone depletion in the stratosphere might lead not only to enhanced photolysis of macromolecules in surface waters ut also to a suppression of the activity of the principal consumers of this photolytically cleaved DOM, and ultimately to an increase in potentially utilizable DOM in the surface layer of the ocean. The pathway of this DOM modified by increased UV-B radiation, however, remains speculative.

Key Message

Bacterioplankton growth rate is a powerful indicator of the decomposition of organic matter and thereby trophic status of the Sea. The decomposition is closely connected to the oxygen consumption in the water column.

The bacterioplankton growth rate represented at least good condition in the off-shore Bothnian Bay and Bothnian Sea. Deep water growth rates were only 50% higher than at corresponding depths in the Atlantic Ocean. The bacterial growth has been stable for the past 13 years, suggesting a balanced nutrient supply to and organic production in the two basins. A tendency of declining bacterioplankton growth, and thereby trophic status, has occurred for the past seven years.

Relevance of the indicator for describing developments of the environment.

Bacterioplankton growth rate is an indicator of the nutrient status in aquatic environments. It is an estimate of the consumption of organic carbon in the ecosystem and therefore closely related to the biochemical oxygen demand in situ (cf. BOD7). Bacterioplankton growth rate thereby indicate the rate of oxygen consumption that may lead to oxygen deficiency in the water column when exceeding oxygen supply. The bacterial growth rate indicator may be used in all aquatic environments.

Bacterioplankton growth rate is a relatively unambiguous indicator of the flux of organic matter in the pelagic ecosystem. Even if the relationship between the factors specific growth rate and abundance may vary, their product representing biomass production reflects the substrate requirement of the bacterial community. Density limitation (i.e. competition) or other limiting factors (i.e. inorganic nutrients, temperature) do not therefore directly appear to control the bacterial community biomass production at typical environmental conditions. This agrees with empirical observations that bacterial growth rate over larger scales correlate with trophic status of a system 1, 2 , and at smaller scales between water layers and seasons 3.

Results

The bacterioplankton growth in the deep water (40-100 m) was stable during 1994-2006 in both the off-shore areas of the Bothnian Bay and Bothnian Sea. No significant linear trends could be demonstrated for the whole period.

A statistically significant decline in bacterioplankton growth could, however, be demonstrated in the northern Bothnian Bay (2000-2006), with a similar tendency at all stations albeit not statistically secure. Significance levels for negative trends at stns. A13 and C1/3 during the same time period were 0,13 and 0,18 , respectively.

Periods of higher and lower growth in the deep water, however, occurred at all stations (c.f. Statistics). The most marked peak occurred at station BMP-A13 during 1997 and BMP-A5 2002.

The bacterioplankton growth was on average 28,5 nmol C dm-3 day-1 in the Bothnian Bay, while being 26% higher in the Bothnian Sea ($p<0.01$, Table 1). The higher value in the Bothnian Sea was in good accordance with the expected higher trophic status of this basin.

The bacterioplankton growth in the Bothnian Sea corresponded to a bacterial oxygen consumption of 1.14 µmol O2 dm-3 day^{-1}, assuming a bacterial growth to growth efficiency function according to del Giorgio and Cole 4. The oxygen consumption could be compared to the average oxygen concentrations of 383 and 267 µmol dm-3, respectively, for the Bothnian Bay and Bothnian Sea. Assuming bacterioplankton to account for half of the total respiration5, average daily oxygen consumption would correspond to 0.51 and 0.85% day $^{-1}$, respectively.

Assessment

The bacterioplankton growth rate in the deep water of the Bothnian Bay and Bothnian Sea was at a level assessed as representing at least good conditions. This quality factor does therefore not support that these basins are disturbed by eutrophication.

The conclusion was based on that the corresponding oxygen consumption was low enough on a daily basis to allow sufficient replenishment of oxygen by advection of surface water. The lack of significant decrease of oxygen levels in both basins since the beginning of the 1970s, corroborate that oxygen supply to the deep water has been and remained sufficient to account for total respiration.

Furthermore, the absolute rates of total oxygen consumption calculated from the bacterial growth (assuming 50% due to bacterial growth) was only 50% higher than the mean oxygen consumption, measured at the same depth interval in the ocean. As elevated nutrient levels are unlikely to occur in the open ocean this further support that the current level of oxygen consumption in the Bothnian Bay and Sea are at the status "good" or better.

The lack of significant long-term and monotonic trends further suggest that the trophic conditions in these basins have been stable for the past decade. Any monotonic trends occurring must be less than 3.5% $year^{-1}$ based on the power of the time series.

9

Sea Water Condition

INTRODUCTION

Most people come in contact with the ocean only near its surface, and usually near its edges. In the huge part of the ocean that remains hidden, sea water is salty, cold, dark, and deep. Average salt content in the ocean is 35 grams per kilogram of sea water, composed mostly of six constituents: sodium (Na+), chloride (Cl-), sulfate (SO_{42}^{-}), magnesium (Mg_2^{+}), calcium (Ca_2^{+}), and potassium (K^+). These are often referred to as conservative elements, because their ratios to each other remain constant throughout the ocean.

It is important to measure salinity of sea water accurately. The salinity and the temperature determine water density (which drives water movement), and concentrations of many elements can be indirectly determined from salinity.

Salinity

In the early twentieth century, salinity was measured by chemically titrating the sample to measure the chloride (Cl^-) ion, then making use of the constant proportions of the major ions to Cl^- to calculate total salinity. In the 1950s and 1960s, it became clear that this procedure and calculation were not satisfactory for the precise measurements needed to distinguish and track different water masses in the oceans.

Drying a sample of sea water and weighing the salt residue is not practical because some of the salts tend to decompose and lose weight before all the water has been removed. The index of refraction of light changes with the salinity of a water sample; a handheld optical instrument called a refractometer can be used to find an inexact measure of salinity. By 1960, the salinity of sea water was most often measured by a salinometer, which measures the electrical conductivity of the sample compared to that of a standard. Electrical conductivity, the ability of a substance to conduct electricity, increases for water as the amount of dissolved ions increase. Using the measurement of conductivity it is possible to measure salinity with high precision, so salinity can be determined to ±0.0001.

The conductivity can be measured in situ (without bringing the water to the surface) at the same time temperature and pressure are measured. These measurements give a profile of salinity and temperature versus depth (pressure) and are commonly collected by an instrument package referred to as a CTD (conductivity, temperature, depth). The units by which salinity is expressed have changed as the methods of measurement changed. Salinity can be given as grams (g) of salt per kilogram (kg) of sea water (g/kg), or as parts per thousand (ppt). Titration for Cl- led to the use of chlorinity units, also expressed as ppt. Sea water of 35 ppt salinity has a 19.4 ppt chlorinity.

Salinity is now defined as the ratio of the electrical conductivity of the sample at 15°C and 1 atmosphere of pressure to that of a potassium chloride (KCl) solution containing 32.4356 g KCl in one kg. of solution. (Atmospheric pressure at sea level is about 14.7 pounds per square inch, termed "1 atmosphere.") The KCl solution is measured at the same temperature and pressure as the sample and gives a conductivity defined to correspond to a practical salinity of 35. Because this is derived from a ratio, it has no units, and is written "salinity is 35." The above definition was formally established in 1980 by the United Nations Educational, Scientific and Cultural Organisation (UNESCO) Joint Panel on Oceanographic Tables and Standards, putting an end to decades of debate about the meaning of salinity. Some authors, uncomfortable with a unitless value, use psu (practical salinity units). In practice, a sample of standard sea water with a salinity of 35.000 is used as a shipboard standard to compare conductivities and calibrate the salinometer.

TEMPERATURE AND DENSITY

The temperature of the world's ocean is highly variable over the surface of the ocean, ranging from less than 0°C (32°F) near the poles to more than 29°C (84°F) in the tropics. It is heated from the surface downward by sunlight, but at depth most of the ocean is very cold. Seventy-five percent of the water in the ocean falls within the temperature range of -1 to +6°C (30 to 43°F) and the salinity range of 34 to 35.

Variations in total salinity and in temperature cause variations in the density of sea water. Several factors can cause the salinity to deviate from 35. Addition of river water or rainwater decreases salinity; excess evaporation or formation of pack ice causes salinity to increase (because ice crystals themselves do not contain salt—the salt is expelled to cracks and pores between the crystals). Cold sea water is denser than warm sea water. There are several areas at the ocean surface where surface water becomes very cold. In these locations, surface sea water becomes denser than the surrounding water and sinks to begin the formation of slow thermohaline currents, which move deep-ocean water. Density differences among different water masses allow physical·oceanographers to calculate movements of water in the ocean. The density of a water sample is a measure of the total mass in a given unit volume.

Salinity increases the density because the dissolved salts are contained in the same volume as the water. Water molecules cluster more closely around positive and negative ions in solution in a process called electrostriction, which also serves to increase sea-water density. Density of water in the ocean, reported as sigma t (st), is calculated from temperature, salinity and pressure by using the equation of state for sea water:st = (s - 1) × 1,000. At 4°C and salinity of 35, the density s of sea water is 1.02781 gram per cubic centimeter, and st = 27.81. At depth, pressure from the overlying ocean water becomes very high (pressure at 4,000 meters is about 400 atmospheres), but water is only slightly compressible, so that there is only a minor pressure effect on density. At a depth of 4,000 metres, water decreases in volume only by 1.8 per cent. Although the high pressure at depth has only a slight effect on the water, it has a much greater effect on easily compressible materials (see box on this page).

DENSITY STRUCTURE OF THE OCEAN

The light and heat from the Sun can only directly penetrate a short distance into the ocean. The surface water is warmer and thus less dense than deep water, which gives most of the ocean a stable density arrangement. The temperature and density often are relatively constant in the surface zone or mixed layer (upper 100 to 200 metres), and begin to change more abruptly in deeper water. The mixed layer is only about two per cent of the total ocean volume, but covers most of its surface.

Temperature decreases and density increases more abruptly in deeper water; the same structure exists in lakes. Many swimmers will recall the experience of finding cooler water at greater depths in lakes. The zone separating surface ocean water from deep water is the pycnocline, containing 18 percent of ocean volume. The deep zone lies below the pycnocline, and contains 80 percent of the ocean volume; its temperature and density are much less variable than those found in the pycnocline.

Most of the ocean lies in complete blackness. Sunlight reaches only depths of about 100 meters (330 feet) in clear open water. This lighted layer is referred to as the photic zone. The depth of light penetration is decreased by particles in the water, including any algal cells that are growing there. Coastal waters with a high sediment content, or water in which an algal bloom is occurring, have much shallower light penetration than clear open-ocean water. Water absorbs different wavelengths of light differently. By a depth of 10 meters (33 feet), mostly blue-green light remains, explaining the bluish color of underwater photos taken in natural light.

Sound

Sound travels at 1,450 metres (4,750 feet) per second in sea water compared to 334 metres (1,100 feet) per second in air.

Sound in water is reflected back when it strikes a solid object. Because the speed of sound in water is well known, this behaviour is used to measure distances under the ocean; a signal is sent out and the time required for the return of the reflected sound can give an accurate measure of the distance to the object that reflected it. This technique is used to measure the depth of water from the surface to the seafloor under a ship; the PDR (precision depth recorder) uses a narrow sound beam to give a continuous record of the water depth through which the ship is moving. Some wavelengths of sound can penetrate the seafloor to some degree, and hence show the layering in sediment. Depth recorders can detect the presence of fish below the surface and record the movements of the deep scattering layer, swarms of small organisms that move toward and away from the sea surface as the time of day changes.

Sonar (sound navigation and ranging) uses sound to locate and identify targets such as submarines. Navies have conducted years of research on sound propagation in water. The speed of sound in water increases as temperature, salinity, and depth increase. Differences in these properties in layers of ocean water cause the sound to refract, or bend, as it travels through the ocean. Refraction can easily make the sound appear to come from a different direction than its real source location, so an accurate understanding of sound physics has been vital for naval operations. A zone of minimum sound velocity exists at a depth of roughly 1,000 meters (about 0.6 mile) called the sofar (sound fixing and ranging) channel. Sound signals that originate in the sofar channel tend to stay in the channel rather than escaping. The sound may travel enormous distances in this channel; explosions set off in the channel in Australia have been heard in Bermuda. A project called ATOC (Acoustic Thermometry of Ocean Climate) was being tested as of 2002. Its goal is to measure global climate change by observing changes in the speed of sound in the sofar channel that would indicate changes in ocean temperatures. Research to determine the effects of this project on marine mammals also is underway.

TRACERS OF OCEAN-WATER MASSES

The oceans, atmosphere, continents and cryosphere are part of Earth's tightly connected climate system. The ocean's role in the climate system involves the transport, sequestration, and exchange of heat, fresh water, and carbon dioxide (CO_2) between the other components of the system. When waters descend below the ocean surface, they carry with them dissolved atmospheric gases. The time-dependent tracers in the oceans provide information on which waters have been in contact with the atmosphere on various timescales. They also give information on the ocean circulation and its variability.

The timescale information is needed to understand and to assess the ocean's role in climate change, and its capacity to take up human-derived constituents, such as CO_2 from the atmosphere. Thus, the advantage to using tracers for ocean circulation studies is the added dimension of time: their time history is fairly well known; they are an integrating quantity; and they provide an independent test for time integration of models and biogeochemical processes. Tracers serve as a "dye" with which to follow the circulation of ocean waters. Conventional ocean tracers include temperature, salinity, oxygen, and nutrients. Stable isotope tracers, such as oxygen-18 and carbon-13, do not decay. In contrast, other radioactive tracers do decay. The radioactive tracers are naturally occurring, such as the uranium/thorium series and radium, and those produced both naturally and by nuclear bomb tests, such as tritium and carbon-14. The bomb contributions from the latter two are called transient tracers, as are the chlorofluorocarbons (CFCs), because they have been in the atmosphere for only a short time. The designation as transient tracer implies a human-derived source and a non-steady input function.

Tracer Timescales Matched to Ocean Processes

The decision of which tracer to use in an oceanographic application depends on the process involved and its timescale. The conventional ocean tracers and stable isotope tracers have no timescale associated with them; their use would depend solely on the natural process involved. For example, to decipher the fresh-water sources in the waters exiting the Arctic Ocean, salinity and the oxygen-18 isotope would be useful. The oxygen isotope is studied because precipitation and the melting and freezing of ice each has different fractions of the ratio of oxygen-18 to oxygen-16.

The radioactive tracers decay with a known half-life. A half-life is the time it takes half of the concentration to disappear. The half-life is matched to what is known about the timescale of the ocean process. For example, to study upper ocean circulation, which occurs on timescales of the order decades, tritium (half-life of 12.4 years) is useful. To study deep ocean circulation, which occurs on timescales of the order hundreds of years, carbon14 (half-life of 5,500 years) is useful. The transient tracers are used to study ocean processes with timescales of less than decades, because that is how long they have existed. This includes upper ocean process, and circulation near the deep-water source regions.

Deep waters that fill ocean basins form primarily in the high latitudes of the North Atlantic and Southern Ocean (the ocean around Antarctica). There is a close coupling of the surface waters in high latitudes to the deep ocean through the density-driven thermohaline circulation. During the process of deep-water formation, atmospheric constituents such as CFCs and CO_2 are introduced into the newly formed water. After these waters sink, they spread out through the deep oceans. As an example of the spreading of deep water from its source in the high latitude North Atlantic, CFCs are used as a tracer.

The Chlorofluorocarbons

The chlorofluorocarbons, CFCs, are synthetic halogenated methanes. Their chemical structures are as follows: CFC-11 is CCl3F; CFC-12 is CCl2F2; and CFC-113 is CCl2FCClF2. The CFCs have received considerable attention because they are a double-edged environmental sword. They are a greenhouse gas and a threat to the ozone layer. The CFCs are used as coolants in refrigerators and air conditioners, as propellants in aerosol spray cans, and as foaming agents. These chemicals were developed in the mid-twentieth century when no one realized they might cause environmental problems. When released, CFCs are gases that have two sinks: the atmosphere, and to a lesser extent, the oceans. Most of the CFCs go up into the troposphere, where they remain for decades. In the ocean and in the troposphere, the CFCs pose no problem.

However, some escape into the stratosphere, where they are a threat to the ozone layer. Destruction of the ozone layer by CFCs removes the atmosphere's ability to block ultraviolet radiation, and has been correlated with the increased incidence of skin cancers. Since the recognition of the CFCs as an environmental problem in the 1970s and the signing of the Montreal Protocol in 1987, the use of CFCs has been phased out. The atmospheric concentrations have started to decrease, as shown in the figure to the right. (The curves represent the amount of CFCs released into the atmosphere [Northern Hemisphere] between 1930 and 2000.) This phase-out is an important international step toward correcting the dangerous trend of stratospheric ozone depletion.

CFCs in the Ocean

The CFCs are gases, and like other gases they get into the ocean via air-sea exchange. There is a direct correlation between gas exchange rate and wind speed, and the direction of the gas flux between the air and ocean is from high to low concentration. For CFCs, the atmospheric concentrations greatly exceed those in the ocean. The concentrations of CFCs dissolved in the surface layer of the ocean depends on the solubility, atmospheric concentration, and other physical factors. Therefore, the colder the water the higher the CFC concentration. The solubility is only slightly dependent on the salinity. The compounds CFC-11 and CFC-12 were first measured in the oceans in the late 1970s. Concentrations generally decrease as the ocean depth increases. An exception is the western North Atlantic, where CFCs reach to the ocean bottom, because of the proximity to the deep-water source and short circulation time.

One of the main advantages to using CFCs as tracers of ocean circulation is that the time-dependent source function permits the calculation of timescales for these processes. A tracer age is the elapsed time since a water parcel was last exposed to the atmosphere. An estimate of age can be calculated from the CFC-11:CFC-12 ratio of the two compounds measured in the oceans and corrected for their solubility. The atmospheric value of the ratio is compared to the atmospheric source function to determine a corresponding date. The date is subtracted from the sample collection date to get an age. Because the atmospheric changes in the ratio of CFC-11:CFC-12 have remained unchanged since the mid1970s, application of the ratio age for CFC-11 and CFC-12 is restricted to older waters. In regions where surface waters are converted to deep and bottom waters which then spread into a background of low-tracer water, a tracer ratio age represents that of the youngest component of the mixture. Thus, the tracers, such as CFCs, can be used to define the pathways, timescale, and transport for the spreading of deep water from its source regions.

10

Marine Ecosystems

INTRODUCTION

Water-borne infection, from marine or freshwater sources, is the leading cause of illness worldwide, and fish provide more animal protein for human consumption than poultry or meat. Today the health of large marine ecosystems—their diversity, productivity, and resilience—is threatened by a "global epidemic" of coastal algal blooms This overgrowth of algae has profound implications for water and food safety because vibrios adhere to phytoplankton (algae) and zooplankton, and "red tides" bear biotoxins responsible for fish and shellfish poisoning. Furthermore nutrient-rich effluents stimulate algal growth and warmer sea surface temperatures shift marine ecosystems towards more toxic species.

Cholera becomes endemic when water and sanitation systems are not kept apart, but other environmental factors affect both the inoculum and persistence of this ancient pathogen. In 1991 cholera struck Peru from Chancay (Lima's port city) to the port of Chimbote 400 km north, the next day. Cholera soon surfaced all along the 2000 km Peruvian coast. It then spread rapidly to ports in Ecuador, Colombia, and Chile, and then to Brazil, Venezuela, and Bolivia, following rivers and streams. Over 15 months, more than half a million people fell ill and almost 5000 died in 19 Latin American nations.

The Peruvian coast is prolific in phytoplankton and their chief consumers, anchovies. Human activities are enhancing the blooms. From March to August, 1991, *Vibrio cholerae* 01, biotype El Tor, serotype Inaba was isolated from marine plankton near Lima; *V cholerae* 01 has been recovered from the bilge of Latin American vessels docked in US Caribbean ports, and seems to have crossed the Pacific as a "stowaway" from Bangladesh.

Since 1960 researchers in Bangladesh have related the seasonality of cholera to coastal algal blooms, but the reservoir remained a mystery. Applying fluorescent antibody and polymerase chain reaction techniques, Colwell and colleagues have identified a viable, but non-culturable "quiescent" form of *V cholerae*, associated with a wide range of surface marine life Under adverse conditions they contract 15-300 fold and reduce their metabolic rates, "hibernating" to tolerate shifts in pH, temperature, salinity, and nutrients. Under favourable conditions of nitrogen, phosphorus, and warming (also conducive to algal blooms), *V cholerae* reverts to a culturable and infectious state. Chitinases and mucinases facilitate attachment to aquatic organisms, while algal surface films and slimes enhance growth by creating turbulence-free havens. Prolonged survival of vibrios has been associated with cyano (blue-green) bacteria, silicated diatoms, drifting dinoflagellates, seaweed, large algae, water hyacinths, and duckweed but the key commensal and/or symbiotic association is with zooplankton. Up to a million bacteria have been detected on copepod (zooplankton) egg sacs *V cholerae* is also found in molluscs and in fish skin and intestines.

HARMFUL ALGAL BLOOMS

In biblical times "red tides" were seen as the blood of sparring whales or as menstrua. Today toxic phytoplankton blooms are associated with paralytic, diarrhoeal, and amnesic shellfish poisoning and with histamine (Scromboid), pufferfish, and ciguatera fish poisoning. The 1972 New England outbreak of red tide paralytic shellfish poisoning, a 1978 outbreak of ciguatera poisoning in Florida (there are now 200 000 cases annually worldwide), and a rash of reports in 1985 on diarrhoeal shellfish poisoning aroused concern, and in 1987 the

International Oceanographic Commission (IOC) and UN Food and Agricultural Organisation (FAO) combined to set up an intergovernmental panel on harmful algae blooms.

Algal blooms (red, green, golden, brown, bioluminescent), covering vast expanses of marine, estuarine, and inland water have now been described from points as diverse as California, North Carolina, Guatemala, Iceland, Japan, Thailand, and the Tasman Sea. The global distribution of coastal blooms is visualised on a composite image from a satellite no longer collecting data; the sea-viewing wide field sensor is scheduled for launching in 1994. This increase in blooms is a direct consequence of human activities, some local and some impacting on global climate systems. The following is just a sample of reports of the consequences for human health, tourism, and food production. In 1973 and 1974 blooms of brevetoxin-producing *Gymnodinium breve* blanketed Florida beaches; in 1976 *G catenatum* bloomed off the Spanish coast causing hundreds of cases of poisoning from contaminated mussels; Pyrodinium blooms off the Philippines in 1983 and 1987 caused 1127 cases and 34 deaths and also 26 deaths on the Pacific coast of Costa Rica and Guatemala in 1987. In the 1980s, diarrhoeal shellfish poisoning associated with dinoflagellates spread from Japan and the Americas to previously uncontaminated areas of Ireland, Portugal, Italy, India and Thailand.

ENVIRONMENTAL FACTORS

An outbreak of waterborne disease may be thought of as an inevitable consequence of specific environmental changes amplifying plankton and associated bacterial proliferation. Sunlight, pH, currents, winds, and river runoffs govern the location and timing of plankton blooms. The major anthropogenic influences, global warming apart, are:

Pollution

Excess nutrient from sewage and fertiliser effluents is a primary cause of marine eutrophication; soil erosion and acid rain add additional nitrogen and phosphorus. Freshwater lacks dissolved carbonates so an increase in atmospheric CO_2 can "fertilise" algal growth on ponds and sewage lagoons. Other pollutants upset the balance of marine ecosystems. Toxins, such as polychlorinated biphenyls, heavy metals, and pesticides, accumulate in food chains, causing damage to marine organisms, altering the ecosystem's equilibrium. Oil slicks and solid plastics harm sea mammals and birds, altering predation pressures.

Over-harvesting

The over-harvesting of fish and shellfish reinforces algal growth also by reducing algivorous grazing. Nine of the world's seventeen major fisheries are in serious decline.

Loss of Habitat

The wetlands ("nature's kidneys") filter nitrogen and phosphorus, store carbon, and support fish and seabirds. They are disappearing. Salt marshes, sea grasses, and mangroves are suffering from coastal urbanisation; and California has now lost most of its wetland area. Coral reefs (the "oceans' rainforests") that protect coasts and cradle marine life are being widely mined for road and housing construction. Warming directly harms reefs, causing the algal symbiont to sprout flagella and swim off, leaving bleached polyps behind.

WARMING AND ALGAL GROWTH

Ship recordings ("sea truth") since 1850 demonstrate ocean warming, and Maskell and colleagues illustrated this in the first article in the series. Confirmation comes from Yan and colleagues' time-series studies in the Western Pacific in the mid-1980s (but the pattern since then has not been consistent. Global warming probably has contributed to recent variations in sea surface temperature but it is not the only possible explanation. The climate phenomenon known as El Niño also raises sea temperatures.

Warming reduces dissolved oxygen and, within limits, stimulates photosynthesis and metabolism, favouring cyanobacteria and dinoflagellates. The 1980s saw several "natural experiments". In 1982/83 a strong El Niño warmed the North Atlantic, altering zooplankton, fish, seal, and seabird communities throughout the decade. In the Pacific raised surface temperatures overwhelmed the cool, rich, upwelling Humboldt current, altering Peruvian sardine and anchovy yields for years.

In 1986/87 warming supported new growths of species in higher northern and southern latitudes. Since 1987 the dinoflagellate *G breve*, previously blooming in the Gulf of Mexico, has persisted off North Carolina, following a shoreward intrusion of the Gulf Stream onto the Continental Shelf. In 1987, local factors combined with warm eddies of the Gulf Stream that swept unusually close to Prince Edward Island, and the pennate diatom (*Nitzschia pungens*) produced domoic acid, killing 5 Canadian mussel consumers and causing 156 cases of amnesic shellfish poisoning. Hundreds of dolphins and whales died that year and there were reverberations throughout the fish and marine mammal populations In September, 1991, domoic acid appeared in Monterey Bay, California, and hundreds of birds were poisoned. In 1992, massive blooms of Pseudonitzschia pseudodelicatissima occurred in Scandinavian waters; and in 1991-92 saxotoxins appeared for the first time as far south as the Straits of Magellan.

In the austral summer the sun beats down on the Pacific to generate an eastward-flowing warming centre known as El Niño. Long cycles in earth's tilt are now bringing the Southern Hemisphere closer to the sun so ever more heat permeates this thermal "sink". Ocean-atmospheric modelling predicts stronger and more frequent El Niños, with rising CO_2. The pace and the duration of strong El Niños, which are associated with extreme weather events worldwide, have accelerated. In 1987-88, 1990-91, and 1992-93 and well into 1993 (and persisting). There have been 2-3° C anomalous sea surface temperature rises associated with floods and droughts. The sea is a global thermostat, absorbing atmospheric heat for worldwide distribution through broad surface "streams" coursing north and grand submarine "rivers" returning to the equator. Might this conveyor belt have stalled or even reversed direction to produce the rapid climate swings disclosed in Greenland ice-core records?

Plankton and Climate

Plankton interact with climate by taking in CO_2, by absorbing and scattering solar heat, and by emitting dimethylsulphide that seeds clouds, which cool the earth's surface through precipitation and sunlight reflection. These biotic feedbacks help to modulate the earth's temperature, the salinity of its oceans, and the composition of its atmosphere.

By harnessing photosynthetically active radiation marine microflora produce twice as much carbohydrate (60 billion tonnes annually) as terrestrial plants. Phytoplankton produce 70% of atmospheric oxygen, that in turn generates protective ozone in the stratosphere. Having evolved before the ozone shield appeared, phytoplankton can produce auxiliary protective pigments, while small increases in the dose of ultraviolet may harm zooplankton (e.g., krill).

EMERGING DISEASES AND NOVEL STRAINS

New clinical conditions (or their potential) involving algae include diarrhoea associated with Cyclospora-like bodies (CLB) and as yet unidentified toxins off the French coast.

While pollution-related stress can increase the susceptibility of sea mammals to infections such as phocine distemper in Mediterranean dolphins it is now clear that vast numbers of viruses exist in marine waters and estuaries. They could be involved in the exchange of genetic information. The role of environmental changes in increasing pathogenicity of marine viruses, or creating conditions for the survival of known pathogenic viruses introduced through sewage may become a crucial question for the health of large marine ecosystems.

In 1992 a modified vibrio known as *V cholerae* 0139 emerged, first in coastal communities of India and now rapidly spreading in Asia; and in sewage lagoons near Lima chlorine-resistant vibrios have been isolated. These developments may not seem easily linked to global climate change but they represent altered biodiversity and stability within large marine ecosystems. Extensive monsoon flooding of Bangladesh in July, 1993, has enhanced the dissemination of the "Bangladesh strains".

CONCLUSION

Climate is controlled by the interaction of oceans, atmosphere, land systems, ice cover, and biota; and a change in any one will destabilise the entire system The transformation in biomass and community structures of key elements in the marine food web is a result of human activities and a byproduct of the ocean's role in the global thermal budget.

The costs in human health and yields are mounting. The degradation of marine ecosystems increases the risk of diseases emerging, and an enhanced plankton reservoir may help explain the rapid invasion of cholera in the Americas. Changes along coastlines are contributing to public health hazards, and are causing hypoxia in the breeding grounds of marine animals and plants. Physicians may see the remedial actions—a reduction in inputs, protection of wetlands, and preservation of species diversity in the oceans—too remote from their clinical practice. Our contribution, and the Lancet series as a whole, is aimed at closing that gap in perception.

11

Dinoflagellate

INTRODUCTION

The dinoflagellates are a large group of flagellate protists. Most are marine plankton, but they are common in fresh water habitats as well. Their populations are distributed depending on temperature, salinity, or depth. About half of all dinoflagellates are photosynthetic, and these make up the largest group of eukaryotic algae aside from the diatoms. Being primary producers make them an important part of the aquatic food chain. Some species, called zooxanthellae, are endosymbionts of marine animals and protozoa, and play an important part in the biology of coral reefs. Other dinoflagellates are colorless predators on other protozoa, and a few forms are parasitic (see for example *Oodinium, Pfiesteria*).

Classification

In 1753 the first modern dinoflagellates were described by Baker and named by Muller in 1773. The term derives from the Greek word *dinos*, meaning 'whirling,' and Latin *flagellum*, a diminutive term for a whip or scourge.

These same dinoflagellates were first defined by Otto Bütschli in 1885 as the flagellate order dinoflagellida. Botanists treated them as a division of algae, named Pyrrhophyta ("fire algae"; Greek *pyrrhos*, fire) after the bioluminscent forms, or

Dinophyta. At various times the cryptomonads, ebriids, and ellobiopsids have been included here, but only the last are now considered close relatives. Dinoflagellates have a known ability to evolve from non-cyst to cyst forming strategies which makes it extremely difficult to recreate their evolutionary history. The dinoflagellat can swim up to 30 miles per hour in saltwater. A dinoflagellate can eat up to 10 times their own bodyweight. Dinoflagellates eat seaweed, sand, other protists or dead fish or sea life.

MORPHOLOGY

Most dinoflagellates are unicellular forms with two *flagella*. One of these extends towards the posterior, called the longitudinal flagellum, while the other forms a lateral circle, called the transverse flagellum. In many forms these are set into grooves, called the *sulcus* and *cingulum*. The transverse flagellum provides most of the force propelling the cell, and often imparts to it a distinctive whirling motion, which is what gives the name dinoflagellate refers to (Greek *dinos*, whirling). The longitudinal acts mainly as the steering wheel, but providing little propulsive force as well.

Dinoflagellates have a complex cell covering called an *amphiesma*, composed of flattened vesicles, called *alveoli*. In some forms, these support overlapping cellulose plates that make up a sort of armor called the theca. These come in various shapes and arrangements, depending on the species and sometimes stage of the dinoflagellate. *Fibrous extrusomes* are also found in many forms. Together with various other structural and genetic details, this organisation indicates a close relationship between the dinoflagellates, *Apicomplexa*, and ciliates, collectively referred to as the alveolates.

The chloroplasts in most photosynthetic dinoflagellates are bound by three membranes, suggesting they were probably derived from some ingested algae, and contain chlorophylls a and c and either peridinin or fucoxanthin, as well as various other accessory pigments. However, a few such as zooxanthellae, which are endosymbionts of marine animals such as sea horses, have chloroplasts with different pigmentation,sexuality, and structure, some of which retain a nucleus. This suggests that chloroplasts were incorporated by several endosymbiotic events involving already colored or secondarily colorless forms. The discovery of plastids in Apicomplexa have led some to suggest they were inherited from an ancestor common to the two groups, but none of the more basal lines have them.

All the same, the dinoflagellate still consists of the more common organelles such as rough and smooth endoplasmic reticulum, Golgi apparatus, mitochondria, lipid and starch grains, and food vacuoles. Some have even been found with light sensitive organelle such as the eyespot or a larger nucleus containing a prominent nucleolus. The dinoflagellate Erythropsidium has the smallest known eye.

LIFE-CYCLE

Dinoflagellates have a peculiar form of nucleus, called a dinokaryon, in which the chromosomes are attached to the nuclear membrane. These lack histones and remain condensed throughout interphase rather than just during mitosis, which is closed and involves a unique external spindle. This sort of nucleus was once considered to be an intermediate between the nucleoid region of prokaryotes and the true nuclei of eukaryotes, and so were termed mesokaryotic, but now are considered advanced rather than primitive traits.

In most dinoflagellates, the nucleus is dinokaryotic throughout the entire life cycle. They are usually haploid, and reproduce primarily through fission, but sexual reproduction also occurs. This takes place by fusion of two individuals to form a zygote, which may remain mobile in typical dinoflagellate fashion or may form a resting dinocyst, which later undergoes meiosis to produce new haploid cells.

However, when conditions become unfavourable, usually when nutrients become depleted or there is insufficient light, some dinoflagellate species alter their life cycle dramatically. Two vegetative cells will fuse together forming a planozygote. Next, is a stage not much different from hibernation called hypnozygote when the organism takes in excess fat and oil. At the same time its shape is getting fatter and the shell gets harder. Sometimes even spikes are formed. When the weather allows it, these dinoflagellates break out of their shell and are in a temporary stage, planomeiocyte, when they quickly reform their individual thecae and return to the dinoflagellates at the beginning of the process. They also move and flagella and some other examples are *Moctiluca* and *Ceratium*.

Ecology

Dinoflagellates sometimes bloom in concentrations of more than a million cells per millilitre. Some species produce neurotoxins, which in such quantities kill fish and accumulate in filter feeders such as shellfish, which in turn may pass them on to people who eat them. This phenomenon is called a red tide, from the color the bloom imparts to the water. Some colorless dinoflagellates may also form toxic blooms, such as *Pfiesteria*. It should be noted that not all dinoflagellate blooms are dangerous. Bluish flickers visible in ocean water at night often come from blooms of bioluminescent dinoflagellates, which emit short flashes of light when disturbed.

Red Tides

The same red tide mentioned above is more specifically produced when dinoflagellates are able to reproduce rapidly and copiously on account of the abundant nutrients in the water. Although the resulting red waves are a miraculous sight, they, again, contain toxins that not only affect all marine life in the ocean but the people who consume them as well. A specific carrier is shellfish. This can introduce both non-fatal and fatal illnesses. One such poison is saxitoxin, a powerful paralytic. Human inputs of phosphate further encourage these red tides, and consequently there is a strong interest in learning more about dinoflagellates, from both medical and economic perspectives.

Evolutionary History

Dinoflagellate cysts are found as microfossils from the Triassic period, and form a major part of the organic-walled marine microflora from the middle Jurassic, through the Cretaceous and Cenozoic to the present day. Because some species are adapted to different surface water conditions, these fossils from sediments can be used to reconstruct past surface ocean conditions. Arpylorus, from the Silurian of North Africa was at one time considered to be a dinoflagellate cyst, but this palynomorph is now considered to be part of the microfauna (Arthropoda). It is possible that some of the Paleozoic acritarchs also represent dinoflagellates.

12

Ciguatera

Ciguatera is a foodborne illness poisoning in humans caused by eating marine species whose flesh is contaminated with a toxin known as ciguatoxin, which is present in many microorganisms (particularly the micro-alga *Gambierdiscus toxicus*) living in tropical waters. Like many naturally and artificially occurring toxins, ciguatoxin bioaccumulates in lower-level organisms, resulting in higher concentration of the toxin at higher levels of the food chain, an example of biomagnification. Predator species near the top of the food chain in tropical waters, such as barracudas, moray eels, parrotfishes, groupers, triggerfishes and amberjacks, are most likely to cause ciguatera poisoning, although many other species have been found to cause occasional outbreaks of toxicity. Ciguatoxin is very heat-resistant, so ciguatoxin-laden fish cannot be detoxified by conventional cooking.

History

Originally, ciguatoxin was linked to poison passed to tropical fish through consumption. However, the exact source of the toxin was unknown, and many sources were identified as the culprit. Some of these included the manchineel fruit, *cocculus berries*, palolo worms, compounds containing copper, pumice, and *corallina opuntia*.

Etymology

It is a generally held theory that ciguatera, as a poisonous substance, was named and identified in Cuba, circa the early 1800s. Local folklore has identified that the etymology stems from a story of an Englishman who caught a barracuda on the Isla de Pinos. After consuming the barracuda, the Englishman became terribly ill. When queried about the origins of his illness, the Englishman claimed to have caught and eaten "a fish, from the seawater". This gave rise to the name of the ailment as ciguatera, a transliteration into Spanish of the English word seawater. Captain Cook during his voyages in the Endeavor (1700s) when off New Caledonia describes eating a fish 'with a large ugly head' and goes on to describe the symptoms of ciguatera poisoning.

Distribution

Due to the localized nature of the ciguatoxin-producing microorganisms, ciguatera illness is only common in tropical waters, particularly the Pacific and Caribbean, and usually is associated with fish caught in tropical reef waters. Ciguatoxin is found in over 400 species of reef fish, and therefore avoidance of consumption of all reef fish (any fish living in warm tropical waters) is the only sure way to avoid exposure to the toxin. Imported fish served in restaurants have been found to contain the toxin and to produce illness which often goes unexplained by physicians unfamiliar with a tropical toxin and its characteristic symptoms. In addition, ciguatoxin has been found in farm-raised salmon.

In February 2008, the FDA reported that several outbreaks of the disease had been traced to fish harvested near the Flower Garden Banks National Marine Sanctuary in the northern Gulf of Mexico, near the Texas-Louisiana shoreline. The FDA advised seafood processors that ciguatera poisoning was "reasonably likely" to occur from consuming any of several species of fish caught as far as 50 miles (80 km) from the sanctuary.

Detection Methods

Currently, multiple laboratory methods are available to detect ciguatoxins, including liquid chromatography-mass spectrometry (LCMS), receptor binding assays (RBA), and neuroblastoma assays (N2A). In Northern Australia, where ciguatera is a common problem, two different methods are widely believed to be available for determining that fish harbors significant levels of ciguatoxin. The first method is that if a piece of fish is contaminated with the toxin, flies will not land on it. The second is that the toxin can be detected by feeding a piece of fish to a cat, as cats are allegedly highly sensitive to ciguatoxin and will display symptoms. It is not known whether there is any veracity to either belief.

Symptoms

Hallmark symptoms of ciguatera include gastrointestinal and neurological effects. Gastrointestinal symptoms include nausea, vomiting, and diarrhea usually followed by neurological symptoms such as headaches, muscle aches, paresthesia, numbness, ataxia, and hallucinations. Severe cases of ciguatera can also result in cold allodynia, which is a burning sensation on contact with cold (commonly incorrectly referred to as reversal of hot/cold temperature sensation). Doctors are often at a loss to explain these symptoms and ciguatera poisoning is frequently misdiagnosed as Multiple Sclerosis.

Dyspareunia and other ciguatera symptoms have developed in otherwise-healthy males and females following sexual intercourse with partners suffering ciguatera poisoning, signifying that the toxin that produces ciguatera poisoning may be sexually transmitted. As diarrhea and facial rashes have been reported in breastfed infants of mothers with ciguatera poisoning, it is likely that ciguatera toxins are also transferred into the breast milk.

The symptoms can last from weeks to years, and in extreme cases as long as 20 years, often leading to long-term disability. Most people do recover slowly over time. Often patients recover but redevelop symptoms in the future. Such relapses can be triggered by consumption of nuts, alcohol, fish or fish-containing products, chicken or eggs, or by exposure to fumes such as those of bleach and other chemicals. Exercise is also a possible trigger.

Treatment

There is no effective treatment or antidote for ciguatera poisoning. The mainstay of treatment is supportive care. Some medications such as the use of Amitriptyline may reduce some symptoms of ciguatera, such as fatigue and paresthesia, although benefit does not occur in every case. Also used are steroids and vitamin supplements, but these merely support the body's recovery rather than directly reducing the toxic effects.

Previously mannitol was used for poisoning after one study reported the reversal of symptoms following its use. Followup studies in animals and case reports in humans also found benefit from mannitol. However, a randomized, controlled, double-blind clinical trial of mannitol for ciguatera poisoning did not find any difference between mannitol and normal saline, and based on this result mannitol is no longer recommended.

There are a number of antiquated Caribbean naturopathic and ritualistic treatments, most of which originated in Cuba and nearby islands. The most common old-time remedy involves bed rest subsequent to a Guanabana juice enema. Other folk treatments range from directly porting and bleeding the gastrointestinal tract to "cleansing" the diseased with a dove during a Santeria ritual. The efficacy of these treatments has never been studied or substantiated; nevertheless they are purportedly still used to this day.

13

Phytoplankton

INTRODUCTION

Phytoplankton are the autotrophic component of the plankton community. The name comes from the Greek words *phyton*, or "plant", and "*planktos*", meaning "wanderer" or "drifter". Most phytoplankton are too small to be individually seen with the unaided eye. However, when present in high enough numbers, they may appear as a green discoloration of the water due to the presence of chlorophyll within their cells (although the actual color may vary with the species of phytoplankton present due to varying levels of chlorophyll or the presence of accessory pigments such as phycobiliproteins, xanthophylls, etc.).

ECOLOGY

Phytoplankton obtain energy through a process called photosynthesis and must therefore live in the well-lit surface layer (termed the euphotic zone) of an ocean, sea, lake, or other body of water. Through photosynthesis, phytoplankton are responsible for much of the oxygen present in the Earth's atmosphere – half of the total amount produced by all plant life. Their cumulative energy fixation in carbon compounds (primary production) is the basis for the vast majority of oceanic and also many freshwater food webs (chemosynthesis is a notable exception). As a side note, one of the more remarkable food chains in the ocean – remarkable because of the small number of links – is that of phytoplankton fed on by krill (a type of shrimp) fed on by baleen whales.

Phytoplankton are also crucially dependent on minerals. These are primarily macronutrients such as nitrate, phosphate or silicic acid, whose availability is governed by the balance between the so-called biological pump and upwelling of deep, nutrient-rich waters. However, across large regions of the World Ocean such as the Southern Ocean, phytoplankton are also limited by the lack of the micronutrient iron. This has led to some scientists advocating iron fertilization as a means to counteract the accumulation of human-produced carbon dioxide (CO2) in the atmosphere. While almost all phytoplankton species are obligate photoautotrophs, there are some that are mixotrophic and other, non-pigmented species that are actually heterotrophic (the latter are often viewed as zooplankton). Of these, the best known are dinoflagellate genera such as Noctiluca and Dinophysis, that obtain organic carbon by ingesting other organisms or detrital material.

The term phytoplankton encompasses all photoautotrophic microorganisms in aquatic food webs. Phytoplankton serve as the base of the aquatic food web, providing an essential ecological function for all aquatic life. However, unlike terrestrial communities, where most autotrophs are plants, phytoplankton are a diverse group, incorporating protistan eukaryotes and both eubacterial and archaebacterial prokaryotes. There are about 5,000 species of marine phytoplankton. There is uncertainty in how such diversity has evolved in an environment where competition for only a few resources would suggest limited potential for niche differentiation. In terms of numbers, the most important groups of phytoplankton include the diatoms, cyanobacteria and dinoflagellates, although many other groups of algae are represented. One group, the coccolithophorids, is responsible (in part) for the release of significant amounts of dimethyl sulfide (DMS) into the atmosphere. DMS is converted to sulfate and these sulfate molecules act as cloud condensation nuclei, increasing general cloud cover. In oligotrophic oceanic regions such as the Sargasso Sea or the South Pacific gyre, phytoplankton is dominated by the small sized cells, called picoplankton, mostly composed of cyanobacteria (Prochlorococcus, Synechococcus) and picoeucaryotes such as *Micromonas*.

14

Aquaculture of Phytoplankton

Phytoplankton are a key food item in both aquaculture and mariculture. Both utilize phytoplankton for the feeding of the animals being farmed. In mariculture, the phytoplankton is naturally occurring and is introduced into enclosures with the normal circulation of seawater. In aquaculture, phytoplankton must be obtained and introduced directly. The plankton can either be collected from a body of water or cultured, though the former method is seldom used. Phytoplankton is used as a foodstock for the production of rotifers, which are in turn used to feed other oganisms. Phytoplankton is also used to feed many varieties of aquacultured molluscs, including pearl oysters and giant clams.

The production of phytoplankton under artificial conditions is itself a form of aquaculture. Phytoplankton is cultured for a variety of purposes, including foodstock for other aquacultured organisms, a nutritional supplement for captive invertebrates in aquaria, and as a source of bio-diesel. Culture sizes range from small-scale laboratory cultures of less than 1L to several tens of thousands of liters for commercial aquaculture. Regardless of the size of the culture, certain conditions must be provided for efficient growth of plankton. The majority of cultured plankton is marine, and seawater of a specific gravity of 1 .010 to 1 .026 may be used as a culture medium. This water must be sterilized, usually by either high temperatures in an autoclave or by exposure to ultraviolet radiation, to prevent biological contamination of the culture. Various fertilizers are added to the culture medium to facilitate the growth of plankton. A culture must be aerated or agitated in some way to keep plankton suspended, as well as to provide dissolved carbon dioxide for photosynthesis. In addition to constant aeration, most cultures are manually mixed or stirred on a regular basis. Light must be provided for the growth of phytoplankton. The colour temperature of illumination should be approximately 6,500 K, but values from 4,000 K to upwards of 20,000 K have been used successfully. The duration of light exposure should be approximately 16 hours daily; this is the most efficient artificial day length.

PHOTOSYNTHETIC PICOPLANKTON

Photosynthetic picoplankton is the fraction of the plankton performing photosynthesis composed by cells between 0.2 and 2 µm (picoplankton). It is especially important in the central oligotrophic regions of the world oceans that have very low concentration of nutrients.

History

- 1952: Description of the first truly picoplanktonic species, Chromulina pusilla, by Butcher This species was renamed in 1960 to Micromonas pusilla and is now known as one of the most abundant in temperate oceanic waters.
- 1979: Discovery of marine Synechococcus by Waterbury and confirmation with electron microscopy by Johnson and Sieburth.
- 1982: The same Johnson and Sieburth demonstrate the importance of small eukaryotes by electron microscopy.
- 1983: W.K. Li and Platt show that a large fraction of marine primary production is due to organisms smaller than 2 µm.
- 1986: Discovery of "prochlorophytes" by Chisholm and Olson in the Sargasso Sea, named in 1992 as Prochlorococcus marinus.
- 1994: Discovery in the Thau lagoon in France of the smallest photosynthetic eukaryote known to date, Ostreococcus tauri, by Courties.

- 2001: Through sequencing of the ribosomal RNA gene extracted from marine samples, several European teams discover that eukaryotic picoplankton are highly diverse.

METHODS OF STUDY

Because of its very small size, picoplankton is difficult to study by classic methods such as optical microscopy. More sophisticated methods are needed.

- Epifluorescence microscopy allows to detect certain groups of cells possessing fluorescent pigments such as Synechococcus which possess phycoerythrin.
- Flow cytometry measures the size ("side scatter") and fluorescence on 1,000 in 10,000 cells per second. It allows one to determine very easily the concentration of the various picoplankton populations on marine samples. Three groups of cells (Prochlorococcus, Synechococcus and picoeucaryotes) can be distinguished. For example Synechococcus is characterized by the double fluorescence of its pigments: orange for phycoerythrin and red for chlorophyll. Flow cytometry also allows to sort out specific populations (for example Synechococcus) in order put them in culture, or to make more detailed analyses.
- Analysis of photosynthetic pigments such as chlorophyll or carotenoids by high precision chromatography (HPLC) allows to determine the various groups of algae present in a sample.
- Molecular biology techniques.
- Cloning and sequencing of genes such as that of ribosomal RNA, which allows to determine total diversity within a sample.
- DGGE (Denaturing Gel Electrophoresis), that is faster than the previous approach allows to have an idea of the global diversity within a sample.
- *In situ* hybridization (FISH) uses fluorescent probes recognizing specific taxon, for example a species, a genus or a class.

- Real-time PCR can be used, as FISH, to determine, the abundance of specific groups. It has the main advantage to allow the rapid analysis of a large number of samples simultaneousl , but requires more sophisticated controls and calibrations.

Composition

Three major groups of organisms constitute photosynthetic picoplankton.

- Cyanobacteria belonging to the genus Synechococcus of a size of 1 µm (micrometre) were first discovered in 1979 by J. Waterbury (Woods Hole Oceanographic Institution). They are quite ubiquist, but most abundant in relatively mesotrophic waters.
- Cyanobacteria belonging to the genus Prochlorococcus are particularly remarkable. With a typical size of 0.6 µm, Prochlorococcus was discovered only in 1988. In spite of its small size, this photosynthetic organism is undoubtedly the most abundant of the planet: indeed its density can reach up to 100 million cells per liter and it can be found down to a depth of 150 m in all the intertropical belt.
- Picoplanktonic eukaryotes are the least well known, as demonstrated by the recent discovery of major groups. Andersen created in 1993 a new class of brown algae, the Pelagophyceae. More surprising still, the discovery in 1994 of a eukaryote of very small size, Ostreococcus tauri, dominating the phytoplanktonic biomass of a French brackish lagoon (étang de Thau), shows that these organisms can also play a major ecological role in coastal environments. In 1999, yet a new class of alga was discovered, the Bolidophyceae, very close genetically of diatoms, but quite different morphologically. At the present time, about 50 species are known belonging to several classes.

Algal classes containing picoplankton species

Classes	Picoplanktonic genera
Chlorophyceae	*Nannochloris*
Prasinophyceae	*Micromonas, Ostreococcus, Pycnococcus*
Prymnesiophyceae	*Imantonia*
Pelagophyceae	*Pelagomonas*
Bolidophyceae	*Bolidomonas*
Dictyochophyceae	*Florenciella*

These approaches implemented since the 1990s for bacteria, were applied to the photosynthetic picoeukaryotes only 10 years later. They revealed a very wide diversity and put in light the importance of the following groups in the picoplankton:

- Prasinophyceae
- Haptophyta
- Cryptophyta

In temperate coastal environment, the genus Micromonas (Prasinophyceae) seems dominant. However, in numerous oceanic environments, the dominant species of eukaryotic picoplankton remain still unknown.

Ecology

Each picoplanktonic population occupies a specific ecological niche in the oceanic environment.

- The Synechococcus cyanobacterium is generally abundant in mesotrophic environments, for example in the vicinity of the equatorial upwelling or in coastal regions.
- The *Prochlorococcus cyanobacterium* replaces it when the waters becomes impoverished in nutrients (i.e. oligotrophic). On the other hand in temperate region (for example in the North Atlantic Ocean),Prochlorococcus is absent because the cold waters prevent its development.

- The diversity of eukaryotes, corresponds undoubtedly to a big variety of environments. In oceanic regions, they are often observed at depth at the base of the well-lit layer (the "euphotic" layer). In coastal regions, certain sorts of picoeukaryotes such as "Micromonas" dominate. Their abundance follows a seasonal cycle, as the plankton of bigger size, with a maximum in summer.

Thirty years ago, it was hypothesized that the speed of division for micro-organisms in central oceanic ecosystems was very slow, of the order of one week or one month. This hypothesis was consolidated by the fact that the biomass (estimated for example by the contents of chlorophyll) was very stable over time. However with the discovery of the picoplankton, it was found that the system was much more dynamic than previously thought. In particular, small predators of a size of a few micrometres which ingest picoplanktonic algae as quickly as they were produced, were found to be ubiquitous. This extremely sophisticated predator-prey system is practically always at equilibrium and results in a quasi-constant picoplankton biomass. This perfect equivalence between production and consumption makes it however extremely difficult to measure precisely the speed at which the system turns over.

In 1988, two American researchers, Carpenter and Chang, had suggested estimating the speed of cell division of phytoplankton by following the course of DNA replication by microscopy. By replacing the microscope by a flow cytometer, it is possible to follow the DNA content of picoplankton cells over time. This allowed to establish that picoplankton cells are extremely synchronous: they replicate their DNA and then divide all at the same time at the end of the day. This synchronization could be due to the presence of an internal biological clock.

Genomics

In the 2000s, genomics allowed to cross a supplementary stage. Genomics consists in determining the complete sequence of genome of an organism and to list every gene present. It is then possible to get an idea of the metabolic capacities of the targeted organisms and understand how it adapts to its environment. To date, the genomes of several types of Prochlorococcus and Synechococcus and of a strain of Ostreococcus have been determined, while those of several other cyanobacteria and of small eukaryotes (Bathycoccus, Micromonas) are under sequencing. In parallel, genome analyses begin to be done directly from oceanic samples (ecogenomics or métagenomics) allowing us to access to large sets of gene for uncultivated organisms.

LIST OF EUKARYOTIC PICOPLANKTON SPECIES

List of eukaryotic species that belong to picoplankton, i.e. which have one of their cell dimensions smaller than 3 µm.

Autotrophic Species

Chlorophyta

Chlorophyceae

- Stichococcus cylindricus (Butcher), 3 - 4.5 µm, brackish

Pedinophyceae

- *Marsupiomonas pelliculata* (Jones et al.), 3 - 3 µm, brackish-marine
- *Resultor micron* (Moestrup), 1.5 - 2.5 µm, marine

Prasinophyceae

- *Bathycoccus prasinos* (Eikrem et Throndsen), 1.5-2.5 µm, marine
- *Crustomastix stigmatica* (Zingone), 3-5 µm, marine
- *Dolichomastix lepidota* (Manton), 2.5-2.5 µm, marine
- *Dolichomastix eurylepidea* (Manton), 3 µm, marine
- *Dolichomastix tenuilepis* (Throndsen et Zingone), 3-4.5 µm, marine
- *Mantoniella squamata* (Desikachary), 3-5 µm, marine
- *Micromonas pusilla* (Manton et Parke), 1-3 µm, marine
- *Ostreococcus tauri* (Courties et Chrétiennot-Dinet), 0.8-1.1 µm, marine

- *Picocystis salinarum* (Lewin), 2-3 µm, hypersaline
- *Prasinococcus capsulatus* (Miyashita et Chihara), 3-5.5 µm, marine
- *Prasinoderma coloniale* (Hasegawa et Chihara), 2.5-5.5 µm, marine
- *Pseudoscourfieldia marina* (Manton), 3-3.5 µm, marine
- *Pycnococcus provasolii* (Guillard), 1.5-4 µm, marine
- *Pyramimonas virginica* (Pennick), 2.7-3.5 µm, marine

Trebouxiophyceae

- *Chlorella nana* (Andreoli et al.), 1.5-3 µm, marine
- *Picochlorum* oklahomensis (Henley et al.), 2-2 µm, hypersaline
- *Picochlorum atomus* (Henley et al.), 2-3 µm, brackish
- *Picochlorum eukaryotum* (Henley et al.), 3-3 µm, marine
- *Picochlorum maculatus* (Henley et al.), 3-3 µm, brackish

Cryptophyta

Cryptophyceae

- *Hillea marina (Butcher)*, 1.5-2.5 µm, marine

Haptophyta

Prymnesiophyceae

- *Chrysochromulina tenuisquama* (Estep et al.), 2-5 µm, marine
- *Chrysochromulina minor* (Parke et Manton), 2.5-7.5 µm, marine
- *Chrysochromulina apheles* (Moestrup et Thomsen), 3 - 4 µm, marine
- *Dicrateria inornata* (Parke), 3-5.5 µm, marine
- *Ericiolus spiculiger* (Thomsen), 3-3.8 µm, marine
- *Imantonia rotunda* (Reynolds), 2-4 µm, marine
- *Phaeocystis cordata* (Zingone), 3-4 µm, marine

- *Phaeocystis pouchetii* (Lagerheim), 3-8 µm, marine
- *Trigonaspis minutissima* (H.A.Thomsen), 2-3.6 µm, marine

Heterokontophyta (Stramenopiles)

Bacillariophyceae

- *Minidiscus comicus* (Takano), 2-7 µm, marine
- *Minidiscus trioculatus* (Hasle), 2.5-3.8 µm, marine
- *Minidiscus spinulosus* (Gao, Chang et Chin), 3-5 µm, marine
- *Minidiscus chilensis* (Rivera), 3-7.5 µm, marine
- *Minutocellus polymorphus* (Hasle, von Stosch et Syverstsen), 2-30 µm, marine
- *Minutocellus scriptus* (Hasle, von Stosch et Syverstsen), 3-36 µm, marine
- *Skeletonema menzelii* (Guillard, Carpenter et Reimann), 2-7 µm, marine
- *Skeletonema pseudocostatum* (Medlin), 2-9 µm, marine
- *Skeletonema japonicum* (Zingone et Sarno), 2-10 µm, marine
- *Skeletonema grethae* (Zingone et Sarno), 2-10.5 µm, marine
- *Skeletonema marinoi* (Sarno et Zingone), 2-12 µm, marine
- *Thalassiosira pseudonana* (Hasle et Heimdal), 2.3-5.5 µm, marine

Bolidophyceae

- *Bolidomonas pacifica* (Guillou et Chrétiennot-Dinet), 1 - 1.7 µm, marine
- *Bolidomonas mediterranea* (Guillou et Chrétiennot-Dinet), 1 - 1.7 µm, marine

Chrysophyceae

- *Ollicola vangoorii* (Conrad (Vors), 2.5-5 µm, marine
- *Tetraparma pelagica* (Booth et Marchant), 2.2-2.8 µm, marine
- *Tetraparma insecta* (Bravo-Sierra et Hernández-Becerril), 2.8-3.8 µm, marine

- *Triparma laevis* (Booth), 2.2-3.1 µm, marine
- *Triparma columacea* (Booth), 2.3-4.7 µm, marine
- *Triparma retinervis* (Booth), 2.7-4.5 µm, marine

Dictyochophyceae

- *Florenciella parvula* (Eikrem), 3-6 µm, marine

Eustigmatophyceae

- *Nannochloropsis granulata* (Karlson et Potter), 2-4 µm, marine
- *Nannochloropsis salina* (Hibberd), 3-4 µm, brackish
- *Nannochloropsis oceanica* (Suda et Miyashita), 3-5 µm, marine

Pelagophyceae

- *Aureococcus anophagefferens* (Hargraves et Sieburth), 1.5-2 µm, marine
- *Aureoumbra lagunensis* (Stockwell et al.), 2.5-5 µm, marine
- *Pelagococcus subviridis* (Norris), 2.5-5.5 µm, marine
- *Pelagomonas calceolata* (Andersen et Saunders), 2-3 µm, marine

Pinguiophyceae

- *Pinguiochrysis pyriformis* (Kawachi), 1-3 µm, marine
- *Pinguiococcus pyrenoidosus* (Andersen et al.), 3-8 µm, marine

Heterotrophic Species

Cercozoa

Cercomonadida

- *Massisteria marina* (Larsen et Patterson), 2.5-6.5 µm, marine

Plasmodiophorea

- *Phagomyxa odontellae* (Kühn, Schnepf & Bulman), 3-4 µm, marine

Stramenopiles

Bicosoecida

- *Caecitellus parvulus* (Patterson et al.), 3-10 µm, marine
- *Pseudobodo minima* (Ruinen), 2 µm, marine
- *Symbiomonas scintillans* (Guillou et Chrétiennot-Dinet), 1.2-1.5 µm, marine

Chrysophyceae

- *Paraphysomonas imperforata* (Lucas), 1.7-5.1 µm, marine
- *Paraphysomonas corbidifera* (Pennick and Clarke), 2-3.25 µm, marine
- *Paraphysomonas antarctica* (Takahashi), 2-4.3 µm, marine
- *Paraphysomonas caelifrica* (Preisig and Hibberd), 2.5-5 µm, marine
- *Paraphysomonas cribosa* (Lucas), 3-4 µm, marine
- *Paraphysomonas sideriophora* (Thomsen), 3-5 µm, marine
- *Paraphysomonas capreolata* (Preisig and Hibberd), 3-6.5 µm, marine
- *Paraphysomonas gladiata* (Preisig and Hibberd), 3-8 µm, marine
- *Picophagus flagellatus* (Guillou et Chrétiennot-Dinet), 1.4-2.5 µm, marine

PLANKTON

Introduction

Plankton consist of any drifting organisms (animals, plants, archaea, or bacteria) that inhabit the pelagic zone of oceans, seas, or bodies of fresh water. Plankton are defined by their ecological niche rather than their genetic classification. They provide a crucial source of food to aquatic life.

Definition

The name plankton is derived from the Greek word "planktos", meaning "wanderer" or "drifter". While some forms of plankton are capable of independent movement and can swim up to several hundreds of meters vertically in a single day (a behaviour called diel vertical migration), their horizontal position is primarily determined by currents in the body of water they inhabit. By definition, organisms classified as plankton are unable to resist ocean currents. This is in contrast to nekton organisms that can swim against the ambient flow of the water environment and control their position (e.g. squid, fish, and marine mammals).

Within the plankton, holoplankton are those organisms that spend their entire life cycle as part of the plankton (e.g. most algae, copepods, salps, and some jellyfish). By contrast, meroplankton are those organisms that are only planktonic for part of their lives (usually the larval stage), and then graduate to either the nekton or a benthic (sea floor) existence. Examples of meroplankton include the larvae of sea urchins, starfish, crustaceans, marine worms, and most fish. Plankton abundance and distribution are strongly dependent on factors such as ambient nutrients concentrations, the physical state of the water column, and the abundance of other plankton.

The study of plankton is termed planktology. Individual plankton are referred to as plankters.

FUNCTIONAL GROUPS

Plankton are primarily divided into broad functional (or trophic level) groups:

- Phytoplankton (from Greek phyton, or plant), autotrophic, prokaryotic or eukaryotic algae that live near the water surface where there is sufficient light to support photosynthesis. Among the more important groups are the diatoms, cyanobacteria and dinoflagellates.
- Zooplankton (from Greek zoon, or animal), small protozoans or metazoans (e.g. crustaceans and other animals) that feed on other plankton and telonemia. Some of the eggs and larvae of larger animals, such as fish, crustaceans, and annelids, are included here.

- Bacterioplankton, bacteria and archaea, which play an important role in remineralising organic material down the water column (note that the prokaryotic phytoplankton are also bacterioplankton).

This scheme divides the plankton community into broad producer, consumer and recycler groups. In reality, the trophic level of some plankton is not straightforward. For example, although most dinoflagellates are either photosynthetic producers or heterotrophic consumers, many species are mixotrophic depending upon their circumstances.

Distribution

Plankton are found in oceans, seas and lakes. However, the local abundance of plankton varies horizontally, vertically and seasonally. The primary cause of this variability is the availability of light. All plankton ecosystems are driven by the input of solar energy (but see chemosynthesis), and this confines primary production to surface waters, and to geographical regions and seasons when light is abundant.

A secondary cause of variability is the availability of nutrients. Although large areas of the tropical and sub-tropical oceans have abundant light, they experience relatively low primary production because of the poor availability of nutrients such as nitrate, phosphate and silicate. This is a result of large-scale ocean circulation and stratification of the water column. In such regions, primary production still usually occurs at greater depth, although at a reduced level (because of reduced light).

Despite significant concentrations of macronutrients, some regions of the ocean are unproductive (so-called HNLC regions). Field studies have found that the mineral micronutrient iron is deficient in these regions, and that adding it can lead to the formation of blooms of many (though not all) kinds of phytoplankton. Iron primarily reaches the ocean through the deposition of atmospheric dust on the sea surface. Paradoxically, oceanic areas adjacent to unproductive, arid regions of continents thus typically have abundant phytoplankton (e.g., the western Atlantic Ocean, where trade winds bring dust from the

Sahara Desert in north Africa). It has been suggested that large-scale "seeding" of the world's oceans with iron could generate blooms of phytoplankton large enough to draw down enough carbon dioxide out of the atmosphere to offset its anthropogenic emissions (responsible for global warming), although other researchers have disputed the scale of this effect.

While plankton are found in the greatest abundance in surface waters, they occur throughout the water column. At depths where no primary production occurs, zooplankton and bacterioplankton instead make use of organic material sinking from the more productive surface waters above. This flux of sinking material can be especially high following the termination of spring blooms.

BIOGEOCHEMICAL SIGNIFICANCE

Aside from representing the bottom few levels of a food chain that leads up to commercially important fisheries, plankton ecosystems play a role in the biogeochemical cycles of many important chemical elements. Of particular contemporary significance is their role in the ocean's carbon cycle.

As stated, phytoplankton fix carbon in sunlit surface waters via photosynthesis. Through (primarily) zooplankton grazing, this carbon enters the planktonic foodweb, where it is either respired to provide metabolic energy, or accumulates as biomass or detritus. As living or dead organic material is typically more dense than seawater it tends to sink, and in open ocean ecosystems away from the coasts this leads to the transport of carbon from surface waters to the deep. This process is known as the biological pump, and is one of the reasons that the oceans constitute the largest carbon sink on Earth.

Some researchers have even proposed that it might be possible to increase the ocean's uptake of carbon dioxide generated through human activities by increasing the production of plankton through fertilization, primarily with the micronutrient iron. However, it is debatable whether this technique is practical at a large scale, and some researchers have drawn attention to possible drawbacks such as ocean anoxia and resultant methanogenesis (caused by the excess production remineralising at depth).

Importance to Fish

Zooplankton are initially the sole prey item for almost all fish larvae as they use up their yolk sacs and switch to external feeding for nutrition. Fish species rely on the density and distribution of zooplankton to coincide with first-feeding larvae for good survival of larvae, which can otherwise starve. Natural factors (e.g. variations in oceanic currents) and man-made factors (e.g. dams on rivers) can strongly affect zooplankton density and distribution, which can in turn strongly affect the larval survival, and therefore breeding success and stock strength, of fish species.

Ocean Acidification

Ocean acidification is the name given to the ongoing decrease in the pH of the Earth's oceans, caused by their uptake of anthropogenic carbon dioxide from the atmosphere Between 1751 and 1994 surface ocean pH is estimated to have decreased from approximately 8.179 to 8.104 (a change of -0.075).

Carbon Cycle

In the natural carbon cycle, the atmospheric concentration of carbon dioxide (CO_2) represents a balance of fluxes between the oceans, terrestrial biosphere and the atmosphere. Human activities such as land-use changes, the combustion of fossil fuels, and the production of cement have led to a new flux of CO2 into the atmosphere. Some of this has remained in the atmosphere (where it is responsible for the rise in atmospheric concentrations), some is believed to have been taken up by terrestrial plants, and some has been absorbed by the oceans. When CO_2 dissolves, it reacts with water to form a balance of ionic and non-ionic chemical species: dissolved free carbon dioxide (CO_2 (aq)), carbonic acid (H_2CO_3), bicarbonate (HCO_3^-) and carbonate (CO_{32}^-). The ratio of these species depends on factors such as seawater temperature and alkalinity.

Acidification

Dissolving CO_2 in seawater also increases the hydrogen ion (H^+) concentration in the ocean, and thus decreases ocean pH.

The use of the term "ocean acidification" to describe this process was introduced in Caldeira and Wickett (2003). Since the industrial revolution began, it is estimated that surface ocean pH has dropped by slightly less than 0.1 units (on the logarithmic scale of pH), and it is estimated that it will drop by a further 0.3-0.5 units by 2100 as the ocean absorbs more anthropogenic CO_2. Note that, although the ocean is acidifying, its pH is still greater than 7 (that of neutral water), so the ocean could also be described as becoming less alkaline.

Although the largest changes are expected in the future, a report from NOAA scientists found large quantities of water undersaturated in aragonite are already upwelling close to the Pacific continental shelf area of North America Continental shelves play an important role in marine ecosystems since most marine organisms live or are spawned there, and though the study only dealt with the area from Vancouver to northern California, the authors suggest that other shelf areas may be experiencing similar effects Similarly, one of the first detailed datasets examining temporal variations in pH at a temperate coastal location found that acidification was occurring at a rate much higher than that previously predicted, with consequences for near-shore benthic ecosystems.

Possible Impacts

Although the natural absorption of CO_2 by the world's oceans helps mitigate the climatic effects of anthropogenic emissions of CO_2, it is believed that the resulting decrease in pH will have negative consequences, primarily for oceanic calcifying organisms. These use the calcite or aragonite polymorphs of calcium carbonate to construct cell coverings or skeletons. Calcifiers span the food chain from autotrophs to heterotrophs and include organisms such as coccolithophores, corals, foraminifera, echinoderms, crustaceans and molluscs.

Under normal conditions, calcite and aragonite are stable in surface waters since the carbonate ion is at supersaturating concentrations. However, as ocean pH falls, so does the concentration of this ion, and when carbonate becomes undersaturated, structures made of calcium carbonate are vulnerable to dissolution. Research has already found that corals coccolithophore algae , coralline algae foraminifera , shellfish nd pteropods experience reduced calcification or enhanced dissolution when exposed to elevated CO_2. The Royal Society of London published a comprehensive overview of ocean acidification, and its potential consequences, in June 2005.

However, some studies have found different response to ocean acidification, with coccolithophore calcification and photosynthesis both increasing under elevated atmospheric pCO_2, an equal decline in primary production and calcification in response to elevated CO_2 or the direction of the response varying between species Recent work examining a sediment core from the North Atlantic found that while the species composition of coccolithophorids has remained unchanged for the industrial period 1780 to 2004, the calcification of coccoliths has increased by up to 40% during the same time.

While the full ecological consequences of these changes in calcification are still uncertain, it appears likely that many calcifying species will be adversely affected. There is also a suggestion that a decline in the coccolithophores may have secondary effects on climate change, by decreasing the earth's albedo via their effects on oceanic cloud cover Aside from calcification, organisms may suffer other adverse effects, either directly as reproductive or physiological effects (e.g. CO_2^- induced acidification of body fluids, known as hypercapnia), or indirectly through negative impacts on food resources However, as with calcification, as yet there is not a full understanding of these processes in marine organisms or ecosystems.

Leaving aside direct biological effects, it is expected that ocean acidification in the future will lead to a significant decrease in the burial of carbonate sediments for several centuries, and even the dissolution of existing carbonate sediments This will cause an elevation of ocean alkalinity, leading to the enhancement of the ocean as a reservoir for CO_2 with moderate (and potentially beneficial) implications for climate change as more CO_2 leaves the atmosphere for the ocean.

15

Protist

INTRODUCTION

Protists are a diverse group of eukaryotic microorganisms. Historically, protists were treated as the kingdom Protista but this group is no longer recognized in modern taxonomy. The protists do not have much in common besides a relatively simple organisation — either they are unicellular, or they are multicellular without specialized tissues. This simple cellular organisation distinguishes the protists from other eukaryotes, such as fungi, animals and plants.

The term *protista* was first used by Ernst Haeckel in 1866. Protists were traditionally subdivided into several groups based on similarities to the "higher" kingdoms: the one-celled animal-like protozoa, the plant-like protophyta (mostly one-celled algae), and the fungus-like slime molds and water molds. Because these groups often overlap, they have been replaced by phylogenetic-based classifications. However, they are still useful as informal names for describing the morphology and ecology of protists.

Protists live in almost any environment that contains liquid water. Many protists, such as the algae, are photosynthetic and are vital primary producers in ecosystems, particularly in the ocean as part of the plankton. Other protists, such as the Kinetoplastids and Apicomplexa are responsible for a range of serious human diseases, such as malaria and sleeping sickness.

CLASSIFICATION

Historical Classifications

The first division of the protists from other organisms came in the 1820's, when the German biologist Georg A. Goldfuss introduced the word protozoa to refer to organisms such as ciliates and corals. This group was expanded in 1845 to include all "unicellular animals", such as Foraminifera and amoebae. The formal taxonomic category Protoctista was first proposed in the early 1860's John Hogg, who argued that the protists should include what he saw as primitive unicellular forms of both plants and animals. He defined the Protoctista as a "fourth kingdom of nature", in addition to the then-traditional kingdoms of plants, animals and minerals The kingdom of minerals was later removed from taxonomy by Ernst Haeckel, leaving plants, animals, and the protists as a "kingdom of primitive forms".

Herbert Copeland resurrected Hogg's label almost a century later, arguing that "Protoctista" literally meant "first established beings", Copeland complained that Haeckel's term protista included anucleated microbes such as bacteria. Copeland's use of the term protoctista did not. In contrast, Copeland's term included nucleated eukaryotes such as diatoms, green algae and fungi. This classification was the basis for Whittaker's later definition of Fungi, Animalia, Plantae and Protista as the four kingdoms of life The kingdom Protista was later modified to separate prokaryotes into the separate kingdom of Monera, leaving the protists as a group of eukaryotic microorganisms. These five kingdoms remained the accepted classification until the development of molecular phylogenetics in the late 20th century, when it became apparent that neither protists or monera were single groups of related organisms (they were not monophyletic groups).

Modern Classifications

Currently, the term protist is used to refer to unicellular eukaryotes that either exist as independent cells, or if they occur in colonies, do not show differentiation into tissues The term protozoa is used to refer to heterotrophic species of protists that do not form filaments. These terms are not used in current taxonomy, and are retained only as convenient ways to refer to these organisms.

The taxonomy of protists is still changing. Newer classifications attempt to present monophyletic groups based on ultrastructure, biochemistry, and genetics. Because the protists as a whole are paraphyletic, such systems often split up or abandon the kingdom, instead treating the protist groups as separate lines of eukaryotes. The recent scheme by Adl et al. (2005) is an example that does not bother with formal ranks (phylum, class, etc.) and instead lists organisms in hierarchical lists. This is intended to make the classification more stable in the long term and easier to update.

Some of the main groups of protists, which may be treated as phyla, are listed in the taxobox at right Many are thought to be monophyletic, though there is still uncertainty. For instance, the excavates are probably not monophyletic and the chromalveolates are probably only monophyletic if the haptophytes and cryptomonads are excluded.

TYPES OF PROTISTS

Protozoa, the Animal-like Protists

Protozoa are mostly single-celled, motile protists that feed by phagocytosis, though there are numerous exceptions. They are usually only 0.01–0.5 mm in size, generally too small to be seen without magnification. Protozoa are grouped by method of locomotion into:

- Flagellates with long flagella e.g., Euglena
- Amoeboids with transient pseudopodia e.g., Amoeba
- Ciliates with multiple, short cilia e.g., Paramecium
- Sporozoa non-mobile parasites; some can form spores e.g., Toxoplasma

Algae, the Plant-like Protists

They include many single-celled organisms that are also considered protozoa, such as Euglena, which many believe have acquired chloroplasts through secondary endosymbiosis. Others are non-motile, and some (called seaweeds) are truly multicellular, including members of the following groups:

- Chlorophytes green algae, are related to higher plants e.g., Ulva
- Rhodophytes red algae e.g., Porphyra
- Heterokontophytes brown algae, diatoms, etc. e.g., Macrocystis

The green and red algae, along with a small group called the glaucophytes, appear to be close relatives of other plants, and so some authors treat them as Plantae despite their simple organisation. Most other types of algae, however, developed separately. They include the haptophytes, cryptomonads, dinoflagellates, euglenids, and chlorarachniophytes, all of which have also been considered protozoans.

Note some protozoa host endosymbiotic algae, as in Paramecium bursaria or radiolarians, that provide them with energy but are not integrated into the cell.

Fungus-like Protists

Various organisms with a protist-level organisation were originally treated as fungi, because they produce sporangia. These include chytrids, slime molds, water molds, and Labyrinthulomycetes. Of these, the chytrids are now known to be related to other fungi and are usually classified with them. The others are now placed among the heterokonts (which have cellulose rather than chitin walls) and the Amoebozoa (which do not have cell walls).

Metabolism

Protists obtain nutrients and digest nutrients in a complex acquirement and assimilation system. Many protists also feed on bacteria, these organisms engulf food and digest it internally. They extend their cell wall and cell membrane around the food material to form a food vacuole. This is then taken into the cell via endocytosis (usually phagocytosis; sometimes pinocytosis).

Nutrition in some different types of protists is variable. In flagellates, for example, filter feeding may sometimes occur where the flagella find the prey.

Nutritional types in protist metabolism

Nutritional type	Source of energy	Source of carbon	Examples
Phototrophs	Sunlight	Organic Compounds or carbon fixation	*Algae, Dinoflagellates* or *Euglena*
Organotrophs	Organic compounds	Organic compounds	*Apicomplexa, Trypanosomes* or *Amoebae*

Reproduction

Some protists reproduce sexually, while others reproduce asexually.

Some species, for example Plasmodium falciparum, have extremely complex life cycles that involve multiple forms of the organism, some of which reproduce sexually and others asexually. However, it is unclear how frequently sexual reproduction causes genetic exchange between different strains of Plasmodium in nature and most populations of parasitic protists may be clonal lines that rarely exchange genes with other members of their species.

- They are eukaryotes because they all have a nucleus.
- Most have mitochondria although some have later lost theirs (Link). Mitochondria were derived from aerobic alpha-proteobacteria that once lived within their cells.
- Many have chloroplasts with which they carry on photosynthesis. Chloroplasts were derived from photosynthetic cyanobacteria living within their cells.
- Many are unicellular and all groups (with one exception) contain some unicellular members.
- The name Protista means "the very first", and some of the 80-odd groups of organisms that we classify as protists may

well have had long, independent evolutionary histories stretching as far back as 2 billion years. But genome analysis added to other criteria show that others are derived from more complex ancestors; that is, are not "primitive" at all.

- Genome analysis also shows that many of the groups placed in the Protista are not at all closely related to one another; that is, the protists do not represent a single clade.
- So we consider them here as a group more for our convenience than as a reflection of close kinship, and
- a better title for this page would be "Eukaryotes that are neither Animals, Fungi, nor Plants".

The Euglenozoa

- Most are unicellular.
- Many swim by means of a single flagellum.
- They are not encased in a cell wall so they are flexible as well as motile.
- Euglena is a typical member of the group (which numbers about 1600 species).
- Because some members of the group (like Euglena) have chloroplasts, these organisms used to be called "Euglenophytes", but in fact they are neither plants ("phytes") nor animals ("zoa").
- Analysis of their genomes suggests, instead, that they are the living descendants of some of the very earliest eukaryotes.
- Trypanosoma brucei, the cause of African sleeping sickness in humans, is a member of the group. The electron micrograph (by L. Tetley; courtesy of Keith Vickerman) shows T. brucei as it occurs in the salivary gland of the tsetse fly ready to be injected into the mammalian host when the fly bites. The specimen is 12 μm long.
- In Latin America, Trypanosoma cruzi, another member of the group, is the cause Chagas disease in humans.

Ciliates, Sporozoans, and Dinoflagellates: the Alveolates

These three phyla are grouped in a clade — the alveolates — because:

- they all have a system of saclike structures ("alveoli") on the inner surface of their plasma membrane as well as;
- close homology in their gene sequences.

Ciliates

- Move by the rhythmic beating of their cilia.
- Although single-celled, some are large enough to be seen with the naked eye.
- Examples: Paramecium, Stentor, Vorticella, Tetrahymena thermophila.
- Feed by sweeping a stream of particle-laden water through a "mouth" and "gullet" and into a food vacuole.
- Undigested wastes are discharged at a permanent site.
- Fresh water ciliates cope with the continuous influx of water from their hypotonic surroundings by pumping it out with one or more contractile vacuoles. Parasitic ciliates, which live in isotonic surroundings, have no contractile vacuole.

All of this rightly suggests that although they are unicellular, there is nothing rudimentary about the ciliates. Their single cell is far more elaborate in its organisation than any cell out of which multicellular organisms are made.

Sporozoans (Apicomplexa)

The members of this group share an "apical complex" of microtubules at one end of the cell (hence the name that many prefer to the old name of sporozoans). All the members of the phylum are parasites.

The genus Plasmodium causes malaria, one of the greatest scourges of humans. Malaria has probably caused more human deaths than any other infectious disease; even today it is estimated to kill a million people a year in the sub-Saharan Africa.

The organism is transmitted from human to human through the bite of mosquitoes of the genus Anopheles.

- The mosquito bite injects sporozoites into the human host.
- These invade the liver where they develop into merozoites.
- The merozoites invade red blood cells where they reproduce.
- Periodically, they all break out of the red cells together bringing on the chills and fever characteristic of the disease.
- Eventually some merozoites develop into either male or female gametocytes.
- These will die unless they are sucked up by the bite of an anopheline mosquito.
- Once in the stomach of the mosquito, the gametocytes form gametes: sperm and eggs.
- These fuse to form zygotes.
- The zygote invades the stomach wall of the mosquito forming thousands of sporozoites.
- These migrate to the salivary gland, ready to be injected into a new human host.

Most forms of malaria are chronic. The organisms may coexist with their host for years (but cannot complete their life cycle there).

Toxoplasma gondii is another parasitic member of this group.

Plasmodium, Toxoplasma, and some of the other members of this group contain a membrane-bounded organelle called the apicoplast. They seem to have inherited it from a common ancestor that acquired it by engulfing a chloroplast. Link to a discussion of this example of secondary endosymbiosis.

Dinoflagellates

- About 1000 species.
- Most are unicellular.
- Most use chlorophylls a and c

- Unlike most eukaryotes, they:
  - lack histones on their chromosomes and
  - have a simpler form of mitosis
- They do have the eukaryotic type ("9 + 2") of flagellum (two of them in fact).
- Occasionally they reproduce explosively, creating poisonous red tides that may cause extensive kills of marine fish and make filter-feeding marine animals like clams unfit for human consumption.

Diatoms, Golden Algae, Brown Algae, and Water Molds

These organisms belong to a single clade (called heterokonts; a/k/a stramenopiles). The first three members share:

- a yellow-brown pigment (which gives them their color). It is a carotenoid called fucoxanthin.
- chlorophylls a and c.

All four of them (plus a number of other groups not listed) share genes closely-homologous to those in red algae. This suggests that they are all descended from a heterotrophic eukaryotic ancestor that acquired a red alga by a secondary endosymbiosis. (While the water molds no longer are photosynthetic, they still retain some red alga genes.)

Diatoms

Diatoms are unicellular. Their cell wall or shell is made of two overlapping halves. These are impregnated with silica and often beautifully ornamented. The photo (courtesy of Turtox) is of Arachnoidiscus ehrenbergi magnified some 400 times.

Diatoms are major producers in aquatic environments; that is, they are responsible for as much as 40% of the photosynthesis that occurs in fresh water and in the oceans. They serve as the main base of the food chains in these habitats, supplying calories to heterotrophic protists and small animals. These, in turn, feed larger animals.

Golden Algae (Chrysophyta)

- Most are unicellular.
- Found in fresh water.
- Important producers in some aquatic food chains.
- In low light conditions, may lose their chlorophyll and turn heterotrophic feeding on bacteria and/or diatoms.
- Over 1000 species alive today; many more in the fossil record.

Brown Algae (Phaeophyta)

- The rockweeds and kelps. Some kelps grow as long as 30 m.
- All are multicellular although without much specialization of cell types.
- Most are found in salt water.
- Used for food in some coastal areas of the world and harvested in the U. S. for fertilizer and as a source of iodine.

Water Molds (Oomycetes)

As their name suggests, water molds were once considered to be fungi. But unlike fungi, the cell wall of water molds is made of cellulose, not chitin. Furthermore, their gene sequences are very different from those of fungi (and most closely related to those of diatoms, golden and brown algae).

Some notable water molds:

- Some species (e.g., Saprolegnia, Achyla) are parasites of fishes and can be a serious problem in fish hatcheries.
- Downy mildews damage grapes and other crops.
- Phytophthora infestans, the cause of the "late blight" of potatoes. In 1845 and again in 1846, it was responsible for the almost total destruction of the potato crop in Ireland. This led to the great Irish famine of 1845-1860. During this period, approximately 1 million people starved to death and many more emigrated to the New World. By the end of the period, death and emigration had reduced the population of Ireland from 9 million to 4 million.

- Phytophthora ramorum, which is currently killing several species of oaks in California.

Red Algae

- The red algae are almost exclusively marine.
- Some are unicellular but most are multicellular.
- Approximately 4000 species have been identified.
- They are photosynthetic using chlorophyll a.
- Their closest relatives are the green algae and land plants.
- However, the structure of the membranes in their chloroplasts is quite different from that of the green plants but resembles that found in the cyanobacteria.
- Like the cyanobacteria, they use:
  - phycoerythrin (which makes them red); and
  - phycocyanin as antenna pigments.
- They do not have the eukaryotic "9+2" flagellum.
- Some are used as food in coastal regions of Asia.
- Agar, the base for culturing bacteria and other microorganisms, is extracted from a red alga.

Slime Molds (Mycetozoa)

Cellular Slime Molds

The organisms in this group have a complex life cycle during the course of which they go through unicellular, multicellular, funguslike (form spores) and protozoanlike (amoeboid) stages.

Thousands of individual amoebalike cells aggregate into a slimy mass — each cell retaining its identity (unlike plasmodial slime molds). The aggregating cells are attracted to each other by the cyclic AMP (cAMP) that they release.

With the exception of one species that causes powdery scab on potatoes, these organisms are of little economic importance.

However, their combination of traits makes them of great scientific interest. The link below will introduce you to one of the most popular members of the group.

Plasmodial (Acellular) Slime Molds (Myxomycetes)

At one stage in their life cycle, these organisms consist of a spreading, slimy, multinucleate mass called a plasmodium that moves slowly over its substrate (e.g., a rotting log) engulfing food and growing as it does so.

Eventually, the plasmodium develops stalks that produce and release spores. If the spores land in a suitable location, they germinate forming single cells that move by both flagella and pseudopodia. These fuse in pairs and start forming a new plasmodium.

Protists without Typical Mitochondria

There are several groups of protists that were long thought to have no mitochondria. However, most (perhaps all) had them in the past. Today, only remnants of their ancestor's mitochondria — called mitosomes — remain.

Some examples are:

❑ *Microsporidia*

- All are unicellular obligate intracellular parasites.
- Many are pathogenic in insects (one is even marketed commercially as a biocontrol agent).
- Some contaminate drinking water supplies and can cause gastrointestinal upsets in humans. Microsporidia, such as *Encephalitozoon cuniculi*, are a common cause of diarrhea in AIDS patients. *Encephalitozoon cuniculi* has a tiny genome [Link] with only 1,997 protein-encoding genes — fewer than many bacteria (e.g., *E. coli* has 4,290). Obliged to live within the cells of its host, it has lost the genes for many important functions (e.g., the citric acid cycle) depending instead on its host.
- Fungi are their closest relatives (not the next two organisms).

- *Giardia intestinalis*
  - Frequently encountered in public water supplies contaminated by animal feces. Causes diarrhea in humans.
- *Entamoeba histolytica.* Causes amebic dysentery, the third most common parasitic disease of humans (after malaria and schistosomiasis).

Choanoflagellates

These are single-celled (e.g., Monosiga), aquatic (both fresh water and marine) protists that have a single flagellum surrounded by a collar ("choano" = collar) of microvilli. Some (e.g., Proterospongia) form simple colonies during part of their life. The flagellum is used for swimming and also beats bacteria-containing water through the collar for feeding.

Sponges also use collar cells to filter food from the water.

Not only does this suggest a close relationship between the two groups, but other evidence indicates that choanoflagellates are the closest protistan relatives of all animals (metazoa). Although single cells, they express genes for several proteins that are essential to cell-cell interactions in metazoans, such as

- cadherins (attach cells to each other — Link)
- tyrosine kinases (used in many examples of cell-cell signaling — Link)

What function these proteins have in the choanoflagellates is unknown.

Other Groups of Eukaryotes

- The metazoa (animals) are described in two separate pages:
  - The Invertebrates; and
  - The Vertebrates.

Index

N

O

❑❑❑